Heat and Mass Transfer Modelling During Drying

Heat and Mass Transfer Modelling During Drying

Empirical to Multiscale Approaches

Mohammad U.H. Joardder
Washim Akram
Azharul Karim

CRC Press is an imprint of the
Taylor & Francis Group, an **informa** business

First edition published 2022
by CRC Press
6000 Broken Sound Parkway NW, Suite 300, Boca Raton, FL 33487-2742

and by CRC Press
2 Park Square, Milton Park, Abingdon, Oxon, OX14 4RN

CRC Press is an imprint of Taylor & Francis Group, LLC

ISBN: 978-1-138-62402-3 (hbk)
ISBN: 978-1-032-05243-4 (pbk)
ISBN: 978-0-429-46104-0 (ebk)

DOI: 10.1201/9780429461040

Typeset in Times
by codeMantra

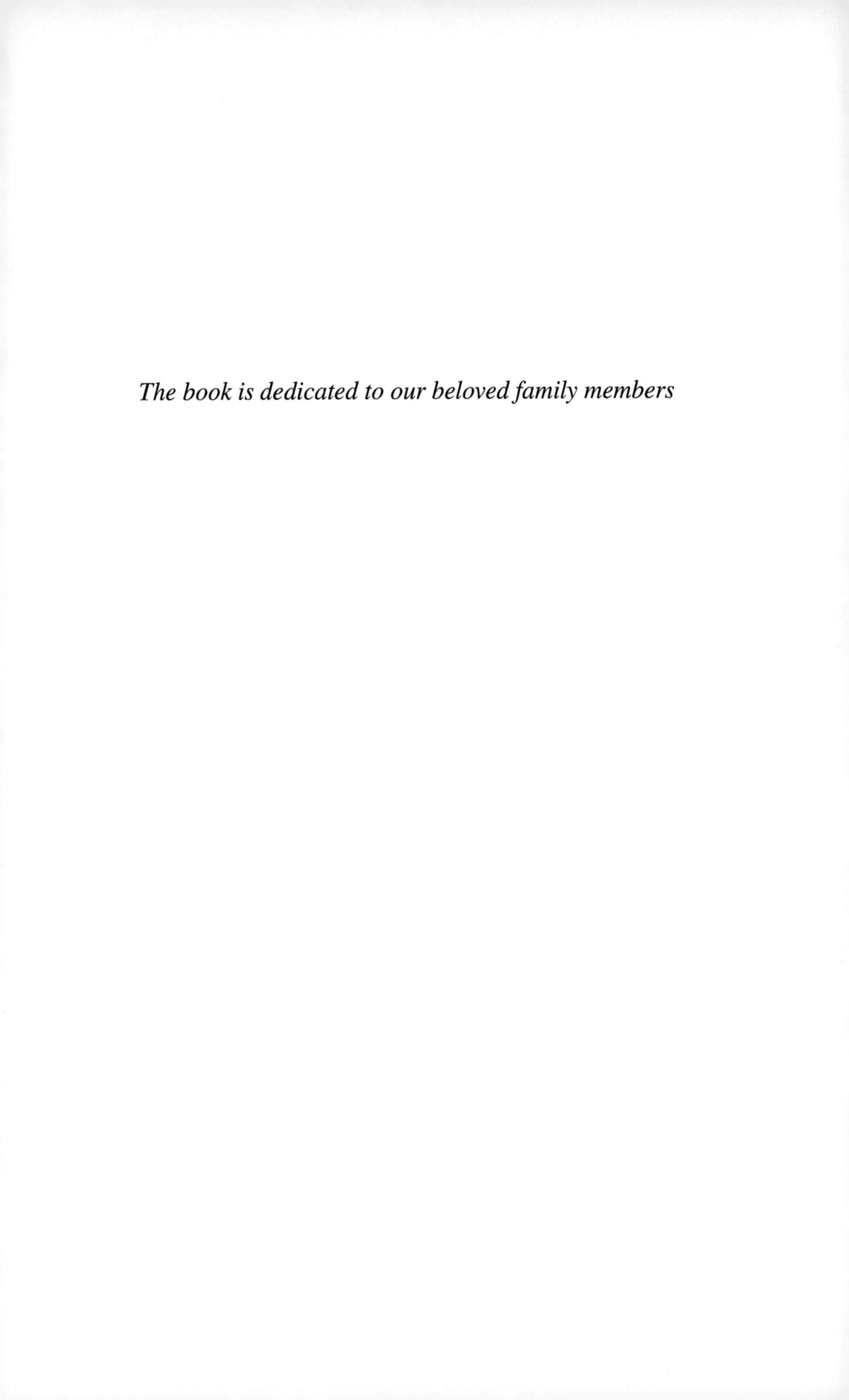

The book is dedicated to our beloved family members

Contents

Preface

An energy-effective dryer needs consideration of energy calculations from the extensive analysis of known physics of drying. Moreover, the quality of dried products is significantly influenced by heat and mass transfer that occur during the drying process.

Most of the conventional dryers use random heating to dry materials without considering their thermal sensitivity and energy requirements. Eventually, excess energy is consumed along with attaining low-quality dried products. Proper heat and mass transfer modelling prior to designing a drying system for selected food materials can overcome these problems. This book extensively discusses this issue of prediction of energy consumption in terms of heat and mass transfer simulation.

A good mathematical model can help to understand the underlying transport phenomena within the materials during drying. However, drying of porous materials such as food is one of the most complex problems in the engineering field that is also multiscale in nature. From the modelling perspective, heat and mass transfer phenomena can be predicted using empirical to multiscale modelling. However, multiscale simulation methods can provide comprehensive insight into the underlying physics of drying food materials.

The objective of this book is to discuss the implementation of different modelling techniques ranging from empirical to multiscale in order to understand the heat and mass transfer phenomena that take place during drying of porous materials such as foods, pharmaceutical products, paper, and leather materials, among others. Relevant properties that are required to develop mathematical modelling approaches will be discussed along with the model development in different approaches. Comparative analysis and feasibility of different modelling approaches will also be included.

The distinct sections discussed in this book are: details of material properties, in-depth discussion on underlying physics of drying, formulation of mathematical modelling from empirical to multiscale approaches, simulation approaches of the mathematical models, possible challenges of different modelling strategies, and options for potential solutions.

Mohammad U.H. Joardder
Washim Akram
Azharul Karim

Acknowledgements

We express our utmost gratitude to the Almighty Creator for His gracious help to accomplish this work. We would like to thank our families for their inspiration and love. Special gratitude goes to Atif Ismaeel, Bint-e-Omar, and Zawjatu Omar for their encouragement and assistance. Also, we would like to express our heartiest gratitude to many people who have supported us in our journey of writing this book.

We sincerely acknowledge the numerous scientific discussions and assistance given by our colleagues in RUET and the members of the Energy and Drying Group at Queensland University of Technology, in particular, Dr. Chandan Kumar, Dr. Mahbubur Rahman, Dr. Md Imran Hossen Khan, Dr. Md. Mahiuddin, and Zachary Welsh.

The authors acknowledge the support from RUET especially the support from the Research Grants Fund of RUET [DRE/6/RUET/249/18 (2018)]. The authors also acknowledge the support from the Advanced Queensland Fellowship (AQF, 02073-2015-30358/15)

We would like to thank Taylor & Francis/CRC Press, the publisher of this book, for giving us the opportunity to share our knowledge and findings with the research community. Special thanks go to Stephen Zollo and Laura Piedrahita, who have keenly worked with us from the inception to completion of this project.

Authors

Dr. Mohammad U.H. Joardder is a Professor in Mechanical Engineering at Rajshahi University of Engineering and Technology (RUET), Bangladesh. He received his PhD degree from Queenslan University of Technology (QUT), Australia in 2016. His research interests include bio-transport, innovative food drying, modelling of novel food processing, food microstructure, and renewable energy. Moreover, his research focuses on applying state-of-the-art computational methods to multiphysics-multiscale transport phenomena and deformation of porous biomaterials. He has authored three popular books with Springer-Nature, three book chapters, and more than 40 peer-reviewed journal publications. Most of his journal articles are in highly ranked journals and have been well cited. He is a regular reviewer of several high-ranked journals of prominent publishers including *Nature*, *Springer*, *Elsevier*, *Wiley*, and *Taylor & Francis*.

Md. Washim Akram completed his BSc in Mechanical Engineering from Rajshahi University of Engineering & Technology (RUET), Bangladesh. He is a faculty member in the Department of Mechanical Engineering at the Bangladesh Army University of Science and Technology (BAUST), Saidpur, Bangladesh. He has published several journal and conference papers. His research interests include drying technology, waste management and energy conversion technology, energy harvesting from renewable sources, and composite materials.

Dr. Azharul Karim is currently working as an Associate Professor in the Mechanical Engineering discipline, Faculty of Engineering, Queensland University of Technology, Australia. He received his PhD degree from Melbourne University in 2007. Dr Karim has authored over 200 peer-reviewed articles, including 113 high-quality journal papers, 13 peer-reviewed book chapters, and 4 books. His papers have attracted about 5,200 citations with *h-index* of 39. His research has had a high impact worldwide, as demonstrated by his overall field weighted citation index (FWCI) of 2.99. He is an editor/board member of six reputed journals including *Drying Technology and Nature Scientific Reports* and supervisor of 26 past and current PhD students. He has been a keynote/distinguished speaker at scores of international conferences and an invited/keynote speaker at seminars in many reputed universities worldwide. He has won multiple international awards for his outstanding contributions in multidisciplinary fields. His research is directed towards solving acute food industry problems by advanced multiscale and multiphase food drying models of cellular water using theoretical/computational and experimental methodologies. Due to the multidisciplinary framework of food drying models, his research spans engineering, mathematics, biology, physics, and chemistry. To address this multidisciplinary challenge, he established the 'Energy and Drying' Research Group consisting of academics and researchers across disciplines.

1 Introduction to Drying

1.1 INTRODUCTION

Drying is one of the oldest preservation methods for perishable materials, including foods, pharmaceutical products, and leather materials. This prehistoric method is also used to extend the shelf life of other materials rich in water content, including wood, paper, and fabrics. Drying is often termed as dehydration as this removes water from the high-moisture materials. Drying offers several advantages including increased shelf life, reduced weight and volume, and reduction of transportation and inventory cost. Although drying is a mature field, there are many issues that still prevail and therefore needs continuous research. Special research and development are essential in many aspects of drying, including the following:

- **Improving the drying kinetics:** Apart from some novel drying methods, most of the drying methods are relatively time-consuming.
- **Minimizing energy consumption**: Drying is an energy-intensive operation and in term of energy, drying is a relatively low energy-efficient process of water removal.
- **Retaining product quality:** Many thermal-sensitive materials, including food and pharmaceutical products, suffer from quality deterioration.

Quality retention, minimizing drying time, and reducing energy consumption are the top priorities of further research in the field of drying. Both drying process conditions and material properties are vital in addressing the abovementioned issues. In this chapter, we discuss several materials that are often subjected to the drying process. Following this, the available drying methods and the optimum drying conditions of these materials will be discussed in detail.

1.2 MATERIALS AND THEIR CHARACTERISTICS

Materials that are subjected to the drying process are diverse in nature. Classification of these materials is not an easy task. However, for the convenience of drying model description, we need to classify the available materials based on their thermophysical and structural characteristics. Figure 1.1 demonstrates the classification of materials in a simplified way.

Solid materials have been conveniently grouped into two basic categories: metallic and non-metallic. Metallic elements including ferrous and non-ferrous items are not subjected to drying in general. Therefore, we will not be discussing

DOI: 10.1201/9780429461040-1

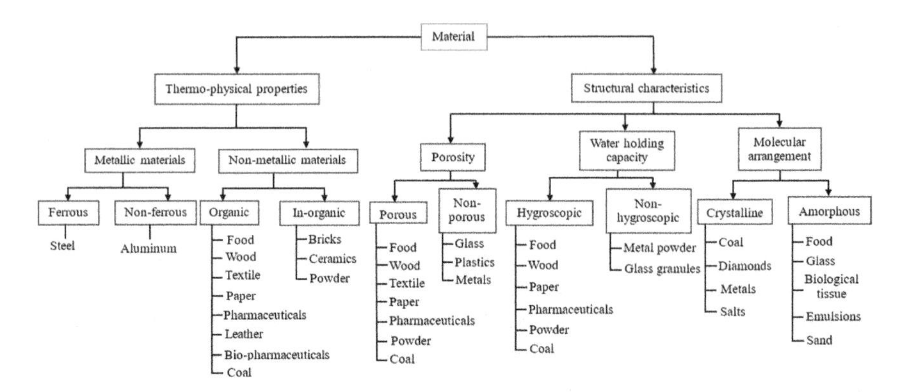

FIGURE 1.1 Classification of materials that are subjected to drying.

metallic samples in this book. A non-metallic substance can be further classified as organic and inorganic.

Organic materials are referred to as those solid substances which consist of carbon compounds and their derivatives of long molecular chains held together with chemical bonds of hydrogen, oxygen, and other materials. Most of the biological materials are composed of carbon compounds and classified as organic solid materials [1]. On the other hand, ceramic, bricks, and powder are the most common examples of inorganic materials. Ceramic materials are generally oxides, nitrides, and carbides of a group of metals including silicon. However, the traditional ceramic encompasses clay minerals, cement, and glass.

Porosity, water-holding pattern, and structural orientation are the common factors that alter the nature of the materials. As these factors significantly affect the thermal properties, they eventually influence the rate of heat and mass transfer and drying kinetics. In this book, we have classified common materials which are subjected to drying on based on these factors.

1.2.1 Porosity

Porosity is defined as the ratio of the volume of pores (containing water and gases) to the total bulk volume of the product. In other words, the ratio of void space to solid materials is called the porosity of the sample. It may range from 0 to 1; however, these are not the practical quantities of most of the materials. Pores can be classified as open pores, closed pores, and blinded pores. Moreover, besides the value of porosity, pore distribution, pore dimension, and pore surface area are also important.

Most of the non-metallic materials possess pores, and the pores may either increase or decrease during drying. Knowledge of porosity is important for the understanding of the nature of density, permeability, and thermal-physical properties; including conductivity and specific heat. Pore size can vary from nanometres to hundreds of micrometres and depending on the size of pores, materials can also be sub-classified as porous or capillary porous.

A material can be defined as capillary porous when its pore size does not exceed 0.1 μm, while porous materials can possess pores bigger than this threshold. Many types of foods, wood, textile, clay, and ceramics can be categorized as capillary porous materials as the size of pores of these materials is generally smaller than 0.1 μm.

1.2.2 Water-Holding Properties

All of the porous materials can retain water to some extent. Based on their water-holding nature, materials can be classified either as hygroscopic or non-hygroscopic. Similar to porosity, hygroscopicity of materials is important in determining permeability and water absorption nature.

A material can be said to be hygroscopic when it possesses a large amount of physically bound water in the solid matrix, as shown in Figure 1.2. However, the definition varies in various fields. The bond of water with a solid matrix determines the degree of hygroscopicity. This can be expressed by the water activity at a certain moisture content of food materials [2]. In the literature, water absorption capacity of the porous surface is defined in various ways, such as hygroscopic moisture. However, in this book, we refer hygroscopic materials to the nature of porous materials that hold water as bound water. Most of the biological materials, including plant-based materials and derived materials from them, are hygroscopic in nature [3]. If the need arises, the hygroscopic nature of the outer surface can be altered by impregnation of non-hygroscopic biodegradable materials, such as starch acetone [4].

Non-hygroscopic materials do not encompass bound water, and the partial pressure of the water in the material is equal to the vapour pressure of the surrounding. Pore spaces in non-hygroscopic materials are filled with water when the material is completely saturated or with air if the materials are completely dry. Moisture transfer during drying in non-hygroscopic materials creates no further

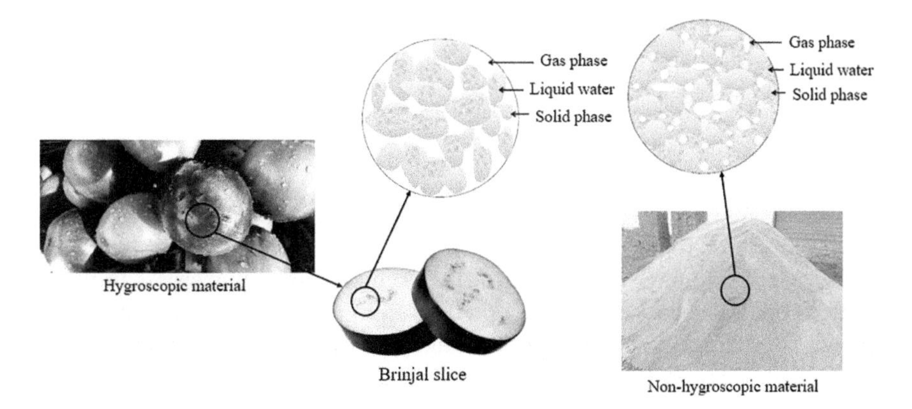

FIGURE 1.2 Hygroscopic (left) and non-hygroscopic (right) materials [3].

complications by introducing deformation. Bricks, ceramics, and pharmaceutical products are often treated as non-hygroscopic materials.

1.2.3 Structural Homogeneity

Materials can be classified as amorphous and crystalline based on the arrangement of atoms or molecules. Crystalline materials have a periodically repeating lattice of atoms or molecules. In this type of material, molecules are packed tightly and are thermodynamically stable. Thus, they are found to be well-shaped. Crystalline nature is less common in non-metallic substances [5].

On the other hand, amorphous solid materials lack the orderly arrangement of atoms or molecules. Most materials that are subjected to drying can be treated as amorphous solids. Moreover, drying techniques including spray drying, hot-air drying, and freeze drying are used to produce amorphous solids. Molecules in the amorphous state are not too tightly packed and possess more pores. Food, wood, leather, paper, and coal can be categorized as amorphous materials.

In the above section, we have defined the physical properties that are important and significantly affect the drying kinetics. Table 1.1 shows a simplified summary of the common characteristics of different materials that are subjected to drying.

It is worth mentioning that the classification of materials is not very straightforward. In various materials, the properties are not so distinct to enable classification. The main concern is the integration of water into the sample as this is influenced by many factors including the abovementioned ones.

In the following section, we will discuss the common materials that are subjected to drying at different stages of their manufacturing, processing, or preservation. All of these materials show the abovementioned characteristics at different degrees. Thus, the drying kinetics vary between these materials.

1.3 COMMON DRYING MATERIALS

Drying drives out water from wet materials to hinder the growth of microorganisms. On the other hand, it is an essential unit process of many materials to reduce their weight and volume.

1.3.1 Food

Accessible free water causes the growth of microorganisms in food materials. Most of raw food materials contain a high amount of water; eventually, these are susceptible to microorganism growth. Drying removes this excess water by simultaneous heat and mass transfer. Apart from prevention of microorganism growth, drying facilitates ease of handling, packaging, shipping, and consumption. Depending on the moisture content and structural characteristics, food can be classified in different ways. However, for the generalization of physio-structural properties, the following classifications are the common ones.

TABLE 1.1
Characteristics of Common Drying Materials

	Types of Common Drying Materials [6–18]									
Characteristics	**Food**	**Wood**	**Textile**	**Paper**	**Pharmaceuticals**	**Leather**	**Bricks and Ceramics**	**Powder Materials**	**Biopharmaceuticals**	**Coal**
Hygroscopic	✓	✓	✓	✓	✓	✓	✓	✓	✓	✓
Non-hygroscopic	–	–	–	–	–	–	✓	✓	✓	–
Porous	✓	✓	✓	✓	✓	✓	✓	✓	✓	✓
Crystalline	–	–	–	–	–	–	✓	✓	–	✓
Amorphous	✓	✓	✓	✓	✓	✓	✓	✓	✓	✓

1.3.1.1 Fruits and Vegetables

Fresh fruits and vegetables are highly perishable as the moisture content in these foods varies between 70% and 90%. Therefore, fruits and vegetables are classified as highly perishable commodities [19]. Either keeping these at lower temperatures or removing the water is the most common approach to extend the stability during storage. It is estimated that about 20%–30% of fruits and vegetables are subjected to drying, whereas more than 50% are consumed as fresh; the remaining are preserved as frozen, canned, and pickled fruits and vegetables [20].

A large amount of fruits and vegetables have been successfully dried using various drying techniques. Some common fruits that are subjected to drying are apple, apricot, fig, peach, pear, and prune raisin. Similarly, beet, carrot, pea, onion, potato, and garlic are important dried vegetables.

1.3.1.2 Grains

Grains contain relatively less moisture content than fresh fruits and vegetables. However, moist grains rapidly deteriorate if stored without drying to the equilibrium moisture content. If the moisture content of the grain is more than the equilibrium moisture content, growth of microorganisms takes place to some extent. The equilibrium moisture content ranges from 13% to 15%, as shown in Table 1.2. For instance, maize needs to be dried to maximum 13.5% moisture content; otherwise, it is susceptible to microorganism growth [21].

1.3.1.3 Leaf and Spices

Green leaves and spices also contain a substantial amount of water in their fresh state. Leaves normally lose up to 85% of their weight during drying [22]. Leaves should be dried carefully as as these are very thin and contain mainly free water and dried leaves have more active nutrient levels than that of fresh leaves [23]. Inappropriate drying results in loss of quality and nutritive constituents of leaves. Therefore, different types of leaves require specific drying conditions to retain the expected quality of the final product.

TABLE 1.2
Equilibrium Moisture Content of Typical Grain

Type of Grain	Maximum Humidity Contents for 1 Year of Storage (or Less) at a Relative Humidity of 70% and a Temperature of 27°C (%)
Maize	13.5
Rice with husk	15.0
Rice without husk	13.0
Sorghum	13.5
Beans	13.5

The main purpose of using spices is the improvement of food flavour and taste owing to the presence of especially volatile components. As the volatile components are very heat-sensitive, low-temperature-based drying should be chosen in the drying of spices.

1.3.1.4 Fish and Meat

Fish and meat contain high moisture contents and are highly perishable foods. In general, fish and meat are preserved using refrigeration. However, refrigeration and freezing are costly options of preservation. Similar to plant-based food materials, meat and fish can be preserved using a drying system for long-term preservation. Fish drying is more common than meat drying.

Due to the intrinsic property of having a wide range of nutritional components, fish is easily seen on the food menu of people all over the world. Fish is replete with vitamins and minerals that can prevent common nutritional deficiencies [24]. People of developed countries usually consume processed fish food at a higher percentage than their developing counterparts. However, due to the lack of proper fish storage or processing technologies, developing countries mainly rely on fresh fish consumption. Contaminating organisms cannot grow in fish with moisture content equal to or below 15%. Meat consists of more fibres than fish, and thus the drying kinetics differs significantly. Apart from this, nutrients present in meat also differ in response to heat during drying. Therefore, drying conditions of fish and meat need to be assessed according to their thermophysical properties and nutrients.

1.3.1.5 Dairy

Fresh milk is extremely perishable without any preservation process; it is not possible to keep milk for even a few hours in consumable condition without some preservation process. One of the long-lasting preservation options is making milk powder from liquid milk. This is not susceptible to microorganism growth, and the quality can be kept as it is in the fresh milk. Using the drying process, water needs to be evaporated to obtain the powder form of milk. However, some novel techniques like membrane techniques can separate a portion of water prior to drying. Spray drying and its advanced alternatives are mainly used in the drying of liquid milk to produce power milk.

1.3.2 Timber

Wood is a fibre-rich material with significantly varied water-holding capacity from its cellular-rich counterparts. Wood encompasses a porous structure consisting of biopolymers along with hydroxyl groups. Cellulose, hemicellulose, and lignin are the main macromolecular components of wood. Structural diversification of wood is prevalent due to the variation of the proportion of the main composition of wood. Figure 1.3 shows the common chemical compositions of wood.

Both free and bound water is presented in wood. Bound water exists mainly in the cell wall of wood, whereas free water can be found in cell lumens and pore.

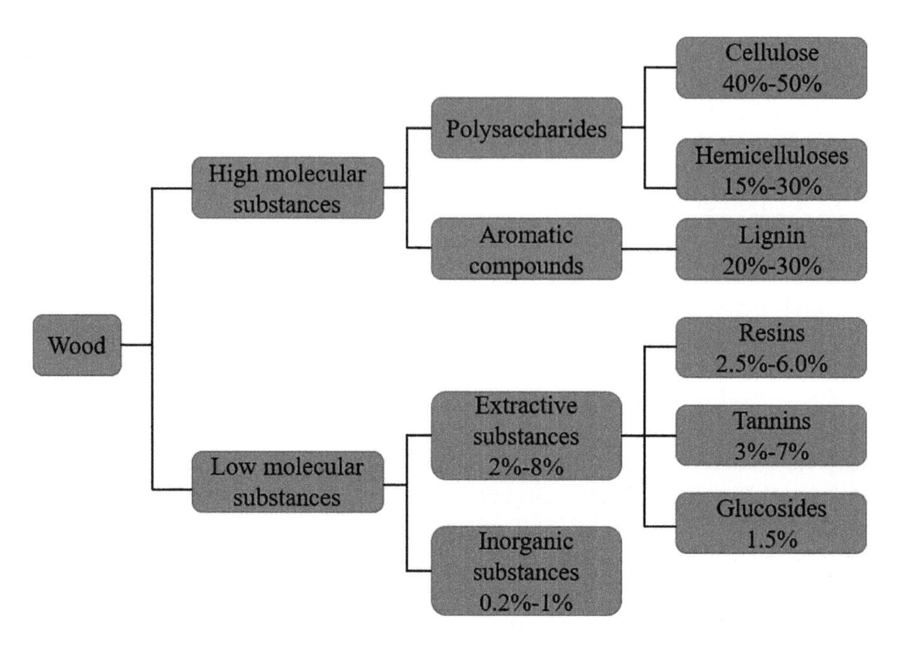

FIGURE 1.3 Chemical composition of wood. (Adapted from [25].)

When all of the free water is removed from a wood sample, the fibre saturation point (FSP) of wood is achieved. The moisture content at FSP varies with types of wood, ranging between 20% and 35% of moisture content. Physical and mechanical properties of FSP of wood significantly depend on moisture content during drying.

1.3.3 Fabrics

Fabrics share many physio-structural properties with wood. Textile is referred to as the material made from fibres, either natural or synthetic. Natural fibres show hydrophilic nature, whereas hydrophobic nature dominates in synthetic fibres. Therefore, the type of fabric influences the water retention capacity of textiles and, thus, affects the drying kinetics. Moisture transfer and distribution patterns also vary with yarn types. Water at the liquid and vapour phases in textile materials is shown in Figure 1.4. In case of natural fibres, moisture transfers through the yarn prior to meeting the fabric surface.

1.3.4 Pulp and Paper

Paper is mainly made of wood and fibre, which are the main components that significantly determine the water-holding capacity of paper. The cellulose fibre is integrated due to hydrogen bonding, and the structural properties are significantly affected by the types of fibre and pulping process. Paper is available

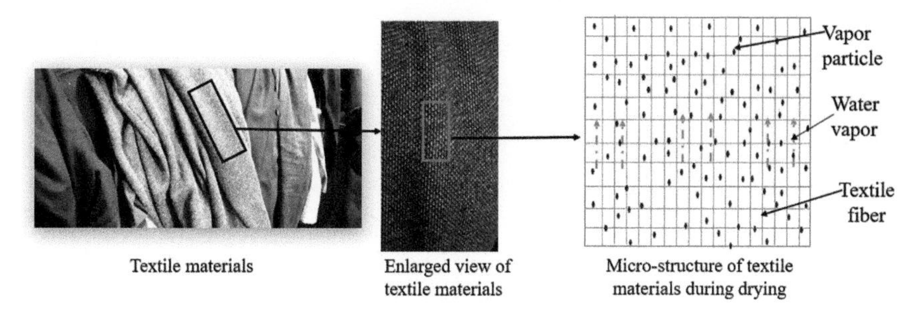

FIGURE 1.4 Water distribution in textile materials [26].

with densities ranging from 610 to 1,150 kg/m³. The thickness of paper generally varies between 60 and 110 μ, and this influences the drying kinetics. Paper shows hygroscopic nature, and the water distribution varies with the degree of this nature.

Drying is one of the essential unit operations for paper producing. It takes enormous energy when compared with the other operations involved in paper production. The pulp of paper has about 90% moisture and 10% fibre. The final moisture content of the paper should be in the range of 6%–9% of its solid contents. Mechanical squeezing is initially used to remove free water, whereas energy-intensive thermal drying removes relatively bound water during the papermaking process.

1.3.5 Chemical and Pharmaceutical Products

Drying is the last unit operation for drugs that are distributed in a non-liquid form, such as tablets and capsules. The initial stage of dried chemicals and pharmaceutical products (CPPs) can be in the form of either granular or paste-like materials or solutions of different compositions. The size of the substance varies between 0.1 μm and 5 mm. Substances used in CPPs, such as antibiotics and fermentation, are thermally sensitive. Low-temperature drying like freeze drying is the most recommended drying system for water removal from CPPs. Vacuum drying is also recommended for many CPPs due to its low-temperature operating conditions. Moreover, an inert gas is used to avoid oxidation in CPP drying.

1.3.6 Leather

Drying is an essential process for achieving the final leather product. Softness is one of the required characteristics of the final leather product. Therefore, choosing drying techniques for leather slightly varies from the selection of drying techniques for other products. Like other biological materials, leather consists of both free and bound water. The micro-level moisture content in leather is shown in Figure 1.5. Drying of leather takes significantly longer, ranging from 36 h to a couple of days.

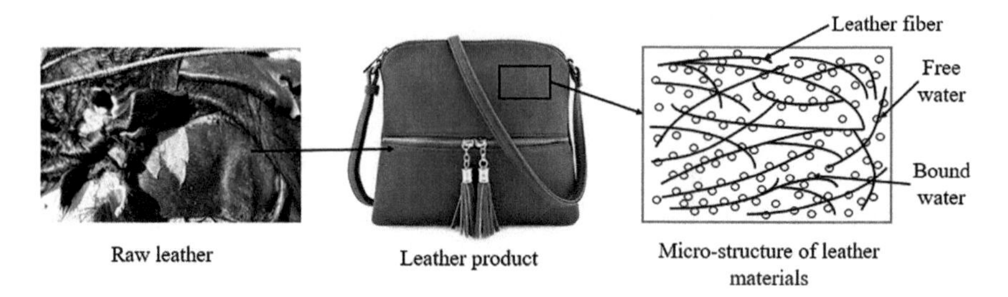

FIGURE 1.5 The micro-level moisture content in leather.

1.3.7 Bricks and Ceramics

Water is an essential element added to hydrate the clay moulding during the processing of bricks and ceramics. Prior to the firing of bricks and ceramics, most of the water needs to be removed from the moulded clay. Excess water in clay products may cause low-quality bricks due to the fast evaporation of water and its consequential deformation. Excess porosity formation or collapse of pores take place in the brick that is not dried prior to the firing operation.

The initial water content of clay products is around 20%–30%, whereas after drying it should contain around 2%–5% of solid materials. Although most of the water in bricks and ceramics is free water, adsorbed water is present, as shown in Figure 1.6, which takes more energy to migrate than capillary water. Brick drying requires between 3,000 and 1,200 kJ/kg of energy depending on the composition and drying conditions. Depending on the water conditions and types of clay material, the usual drying period of clay is 3–10 days.

1.3.8 Coal

Coal is used as a solid fuel in the different process, including electricity production. Drying of coal increases the calorific value and decreases the transportation cost. Water is present with different bonding strengths with the solid matrix and can be classified as free, physically bound, and chemically bound water, as shown in Figure 1.7; around 80% of water present in coal is free water.

The expected moisture content in coal varies with the different processing and application methods, and it ranges from 0% to 15%. Coal matrix encompasses

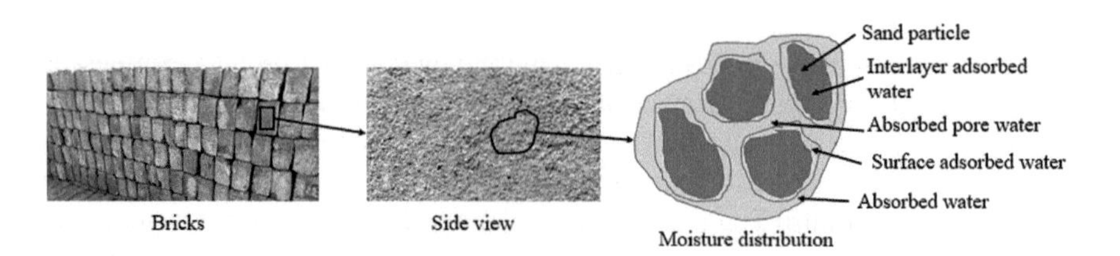

FIGURE 1.6 Moisture distribution in bricks.

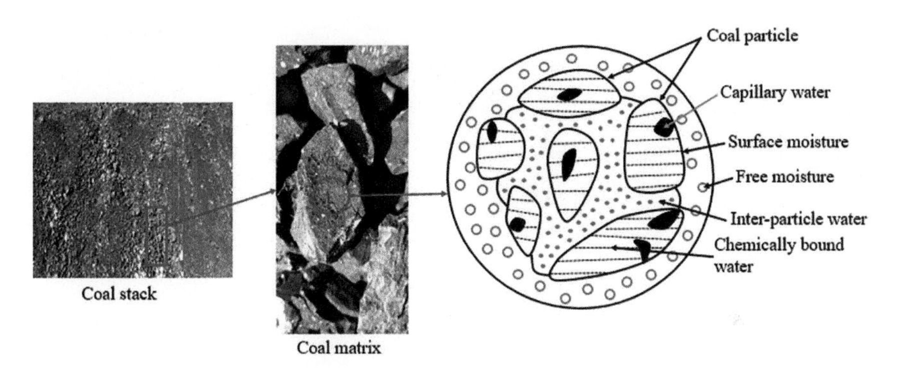

FIGURE 1.7 Forms of moisture related to coal. (Adapted from [27].)

both aromatic and hydro-aromatic components, holding moisture and other volatile components. Different grades of pores, namely, micro-pores, mesopores, and macropores, are seen in coals. Depending on the properties, moisture content, and heating values, coals are classified as high and low ranks.

Unlike the drying of other materials mentioned in the earlier sections, hazards including fire, explosion, and decomposition are associated with coal drying. Therefore, special safety measures must be taken during coal drying to avoid personnel injury.

1.4 DRYING PHENOMENA AND METHODS

Drying is mainly carried out to remove water from the sample, and it is an essential unit operation in the food, chemical, polymer, pulp and paper, ceramics, pharmaceutical, and wood processing industries. In general, typical drying process can be divided into constant rate period and falling rates periods.

In the constant rate stage of drying, only free water effectively migrates from wet materials. In this period, the evaporation occurs from the surface at a constant rate and the internal moisture transfer mechanism towards the surface is sufficient enough to balance the surface equilibrium. As the internal moisture transport to the externally exposed surface is equal to the evaporation rate, the energy input and heat loss during dehydration remain constant.

After critical moisture content is reached, materials are in the falling rate stage of drying and bound water is removed during this period. The falling rate period is subcategorized as first falling rate and second falling rate periods. The internal moisture flow rate falls in the first falling stage; eventually, the evaporation rate decreases as the heat flux from the surface increases.

On the other hand, relatively more time is required in the second falling stage than the first one. The second falling period of drying needs roughly the same energy as the first falling rate period to remove the last 10% moisture. However, this figure varies based on the materials being dried and the drying technique followed.

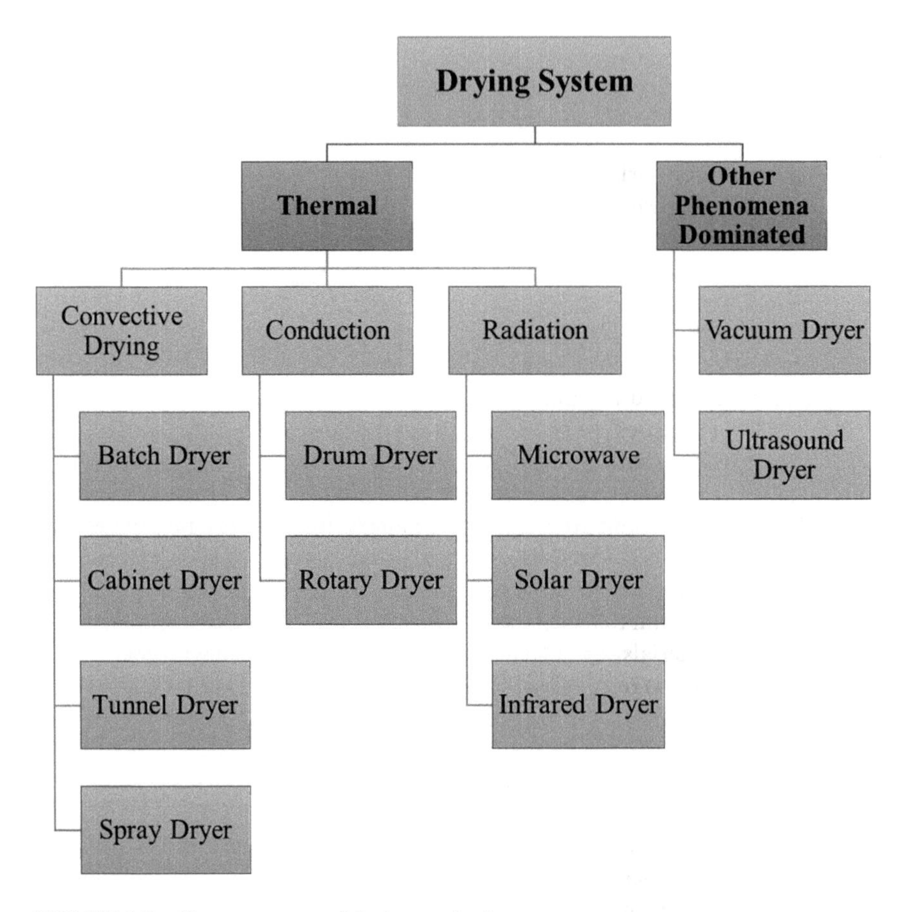

FIGURE 1.8 Common types of drying methods.

1.4.1 Common Drying Methods

There are over 300 types of dryers that have been reported in the literature; however, some common drying systems, as demonstrated in Figure 1.8, dominate the drying industries. Process conditions such as temperature, pressure, and the humidity of the drying system vary significantly in the different drying systems. Energy consumption for industrial drying operations accounts for approximately 10%–25% [28].

Some common drying types are briefly described in the following section. The readers interested in more insight into different types of drying can consult numerous other sources available, such as those listed in References [29,30].

1.4.1.1 Convective Drying

Convection drying is the most common drying method. Convectional hot-air drying is associated with convection heat transfer from the air to a wet surface, followed by conduction heat transfer towards the sample core. Due to the low heat and moisture diffusivity of porous high-moisture materials, water removal from sample cores to the surrounding takes an exceptionally long time. It takes even longer time to remove bound water from the food materials [31]. Moreover, case hardening phenomena occurs during convective drying due to the continuous heating of the food surface. Depending on the source of heat supplied to the air, this type of drying can include sun drying, heat pump drying, and other drying systems.

1.4.1.2 Microwave Drying

Microwave (MW) does not fall in the range of thermal radiation in the electromagnetic spectrum. In this, a direct incident MW falling on an object does not result in increased thermal energy. In practice, microwave interaction with dipole molecules causes tremendous motion, producing heat energy much faster [32]. MW penetrates sufficiently deep into the wet sample, resulting in volumetric heating [33].

MW drying takes advantages of the response of water molecule to the electromagnetic field with a wavelength of 1 m–1 mm [34]. Microwaves are generated by gradually increasing the frequency from 60 Hz to 2,450 MHz, in a stepwise manner, using vacuum tube devices such as magnetron.

In the wet materials, water molecules absorb MWs, causing random motion of water molecules to result in heat generation and eventually transfer heat to the rest of the components of the moisture-rich sample. MW volumetric heating takes place in the sample in which water is available, such as in food material. MW heating can remove various kinds of water, including physically bound water.

1.4.1.3 Infrared Drying

Infrared (IR) dryer uses the radiation mode of heat transfer. Radiation heat transfer governs the moisture migration from the surface of the sample without assistance from any other medium. Therefore, less thermal resistance is observed during infrared drying for the sample with higher emissivity. In the absence of hot air, direct contact of oxygen with the sample material can be avoided in infrared drying, thus assisting in retaining the high quality of oxidation-prone samples. Due to the intrinsic advantages of radiation heat transfer, IR offers easy removal of bound water removal, saves energy, and has a short drying time.

1.4.1.4 Vacuum Drying

Vacuum drying is one of the non-thermal drying techniques that is mainly suitable for the drying of heat-sensitive materials including foods and pharmaceutical products [35]. Vacuum pressure is developed in vacuum drying, where water can evaporate at room temperature. The vacuum in the drying chamber has a twofold function on the products being processed; one is to reduce the surface water content of the products and the other is to reduce the boiling point of the inner water content. Hence, a higher vapour pressure gradient can be observed between the interior and the surface of the product, eventually increasing the drying rate.

1.4.1.5 Freeze Drying

Freeze drying is one of the best drying methods for the products that are heat sensitive. Freeze drying involves removal of the ice crystals from frozen foods through sublimation. The pressure in freeze drying is maintained in such a fashion that it keeps the sublimating ice pressure as well as the pressure required for the direct transformation of solid ice to water vapour below the triple point of water. The first stage of the freeze-drying process starts with the quick freezing of raw materials at –18°C. The second stage of freeze drying is sublimation, which causes the removal of free water from the raw products through air drying. The frozen water in the products is drained out in the form of steam under very low-pressure conditions. The heat energy required to carry out the sublimation process can be obtained either from the temperature gradient between the vacuum chamber and products or using the integrated heating chamber. To ease the flow of water vapour towards the condenser, the drying chamber as well as the condenser can be maintained in vacuum conditions itself. After passing the condenser area, the steam of water is accumulated as ice.

1.4.1.6 Spray Drying

All of the above-mentioned drying techniques are used to remove water from porous solid materials. Spray drying, on the other hand, is a good choice for powder production from a liquid counterpart. Spray drying is the process of converting the fluid solution flow into the dry particulate form in the presence of a hot medium. The drying temperature is much higher than that for conventional hot-air drying and ranges from 150°C to 200°C. The final form of the product may be either granular, in powder form, or as agglomerates, depending on the composition and flow pattern. A spray-drying process consists of the atomization of the feed sample, liquid sample drying, and recovery of powder.

The liquid is fed to the drying chamber after atomization through an atomizer that atomizes the liquid into tiny, fine droplets with the maximum exposed surface area for efficient drying. After evaporation of moisture from the droplets, two stages of separation take place. The powder of the raw liquid is eventually collected after these separation stages.

1.4.2 Drying Conditions

Diverse drying conditions are associated with different types of drying systems. However, the main purpose of the drying parameters is developing an environment that assists in removing water from wet porous materials, in other words, providing any form of energy to the water molecules to overcome the latent heat of evaporation [36]. Most of the drying involves the application of heat in different modes, namely, convection, conduction, and radiation. About 85% of industrial dryers are of the convective type, with hot air as the drying medium. Drying parameters for hot-air drying are air temperature, air velocity, and humidity. On the other hand, vacuum pressure is the only important parameter for vacuum drying. Optimum drying conditions of different materials are shown in Figure 1.9.

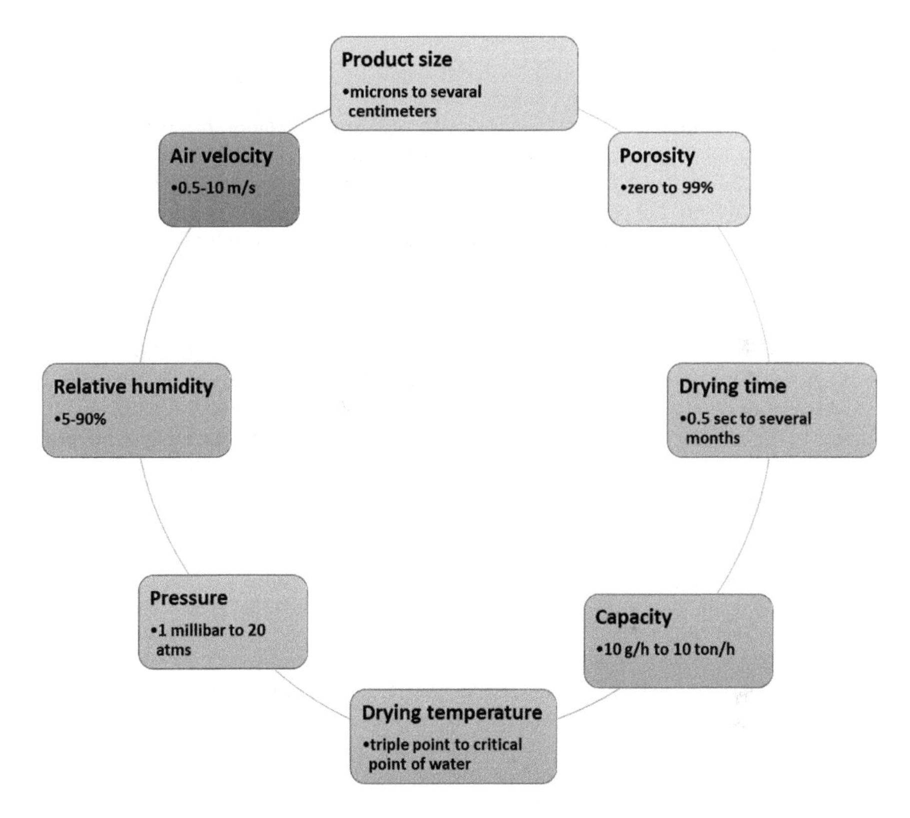

FIGURE 1.9 Optimum drying conditions of different materials. (Data are compiled from several sources including [14,37–47].)

From the above figure, it can be seen that drying conditions have a wide range of individual parameters that need to be taken care of, due to the nature of moisture-rich materials. For example, air velocity can be 0.5–10 m/s for different materials. Light weight samples such as leaves need slower air velocity during drying, whereas larger samples can withstand the high velocity of air during drying. The drying conditions also vary due to the scope of drying. In summary, drying conditions need to be maintained according to the characteristics of wet materials.

REFERENCES

1. M. Mahiuddin, M.I.H. Khan, N. Duc Pham, M.A. Karim, "Development of fractional viscoelastic model for characterizing viscoelastic properties of food material during drying," *Food Biosci.*, vol. 23, pp. 45–53, 2018
2. M. U. H. Joardder, M. Mourshed, and M. H. Masud, *State of Bound Water: Measurement and Significance in Food Processing*. Springer, 2019.
3. M. U. H. Joardder, A. Karim, C. Kumar, and R. J. Brown, *Porosity: Establishing the Relationship between Drying Parameters and Dried Food Quality*. Springer, 2015.

4. F. D. S. Larotonda, K. N. Matsui, P. J. A. Sobral, and J. B. Laurindo, "Hygroscopicity and water vapor permeability of Kraft paper impregnated with starch acetate," *J. Food Eng.*, vol. 71, no. 4, pp. 394–402, 2005.
5. M. U. H. Joardder, M. H. Masud, and M. Azharul, "Relationship between intermittency of drying, microstructural changes, and food quality," in *Intermittent Nonstationary Dry. Technol. Princ. Appl.*, Ed., C. L. Law and A. Karim. CRC Press, 2017, pp. 123–137.
6. R. Li, D. Lin, Y. H. Roos, and S. Miao, "Glass transition, structural relaxation and stability of spray-dried amorphous food solids: A review," *Dry. Technol.*, vol. 37, no. 3, pp. 287–300, 2019.
7. F. O. Tesoro, E. T. Choong, and O. K. Kimbler, "Relative permeability and the gross pore structure of wood," *Wood Fiber Sci.*, vol. 6, no. 3, pp. 226–236, 2007.
8. H. Derluyn, H. Janssen, J. Diepens, D. Derome, and J. Carmeliet, "Hygroscopic behavior of paper and books," *J. Build. Phys.*, vol. 31, no. 1, pp. 9–34, 2007.
9. V. Murikipudi, P. Gupta, and V. Sihorkar, "Efficient throughput method for hygroscopicity classification of active and inactive pharmaceutical ingredients by water vapor sorption analysis," *Pharm. Dev. Technol.*, vol. 18, no. 2, pp. 348–358, 2013.
10. R. Molina, E. Clemente, M. R. da S. Scapim, and J. M. Vagula, "Physical evaluation and hygroscopic behavior of dragon fruit (Hylocereus undatus) lyophilized pulp powder," *Dry. Technol.*, vol. 32, no. 16, pp. 2005–2011, 2014.
11. S. V. Jangam, M. Karthikeyan, and A. S. Mujumdar, "A critical assessment of industrial coal drying technologies: Role of energy, emissions, risk and sustainability," *Dry. Technol.*, vol. 29, no. 4, pp. 395–407, 2011.
12. M. M. Rahman, M. U. H. Joardder, M. I. H. Khan, N. D. Pham, and M. A. Karim, "Multi-scale model of food drying: Current status and challenges," *Crit. Rev. Food Sci. Nutr.*, vol. 58, no. 5, pp. 858–876, 2018.
13. R. D. Buck, "A note on the effect of age on the hygroscopic behaviour of wood," *Stud. Conserv.*, vol. 1, no. 1, pp. 39–44, 1952.
14. L. H. C. D. Sousa, O. C. Motta Lima, and N. C. Pereira, "Analysis of drying kinetics and moisture distribution in convective textile fabric drying," *Dry. Technol.*, vol. 24, no. 4, pp. 485–497, 2006.
15. J. Monzó-Cabrera, J. M. Catalá-Civera, P. Plaza-González, and D. Sánchez-Hernández, "A model for microwave-assisted drying of leather: Development and validation," *J. Microw. Power Electromagn. Energy*, vol. 39, no. 1, pp. 53–64, 2004.
16. M. Ameri and Y.-F. Maa, "Spray drying of biopharmaceuticals: Stability and process considerations," *Dry. Technol.*, vol. 24, no. 6, pp. 763–768, 2006.
17. M. Le Roux, Q. P. Campbell, M. J. Van Rensburg, E. S. Peters, and C. Stiglingh, "Air drying of fine coal in a fluidized bed," *J. South. African Inst. Min. Metall.*, vol. 115, no. 4, pp. 335–338, 2015.
18. M. J. Van Rensburg, M. Le Roux, Q. P. Campbell, and E. S. Peters, "Moisture transport during contact sorption drying of coal fines," *Int. J. Coal Prep. Util.*, vol. 40, no. 4–5, pp. 281–296, 2018.
19. M. U. H. Joardder and M. H. Masud, *Food Preservation in Developing Countries: Challenges and Solutions*. Springer, 2019.
20. M. I. H. Khan and M. A. Karim, "Cellular water distribution, transport, and its investigation methods for plant-based food material," *Food Res. Int.*, vol. 99, pp. 1–14, 2017.
21. "Drying the grain its importance and common practices," pp. 1–10. Retrieved May 20, 2021, from https://www.shareweb.ch/site/Agriculture-and-Food Security/focusareas/Documents

22. G. Chen and A. Mujumdar, "Drying of herbal medicines and tea," in *Handbook of Industrial Drying*, Ed. A. S. Mujumdar, Dekker, 3rd edition, 2006.
23. S. R. Navale, U. Supriya, V. M. Harpale, and K. C. Mohite, "Effect of solar drying on the nutritive value of fenugreek leaves," *Int. J. Eng. Adv. Technol.*, vol. 4, no. 2, pp. 2249–8958, 2014.
24. FAO, *The State of World Fisheries and Aquaculture 2018—Meeting the Sustainable Development Goals.* Rome, 2018.
25. C. M. Popescu, "Wood as bio-based building material," in *Performance of Bio-based Building Materials*, Ed. D. Jones and C. Brischke. Woodhead, 2017, pp. 21–96.
26. J. Ter Schiphorst, M. Van Den Broek, T. De Koning, J. N. Murphy, A. Schenning, and A. C. C. Esteves, "Dual light and temperature responsive cotton fabric functionalized with a surface-grafted spiropyran–NIPAAm-hydrogel," *J. Mater. Chem. A*, vol. 4, no. 22, pp. 8676–8681, 2016.
27. M. Karthikeyan, W. Zhonghua, and A. S. Mujumdar, "Low-rank coal drying technologies—Current status and new developments," *Dry. Technol.*, vol. 27, no. 3, pp. 403–415, 2009.
28. M. F. Ashby and D. R. H. Jones, *Engineering Materials 1: An Introduction to Properties, Applications and Design*, vol. 1, Elsevier, 2012.
29. A. S. Mujumdar, "Principles, classification, and selection of dryers," in *Handbook of Industrial Drying*, CRC Press, 2006, pp. 28–57.
30. M. M. Rahman, M. U. H. Joardder, and A. Karim, "Non-destructive investigation of cellular level moisture distribution and morphological changes during drying of a plant-based food material," *Biosyst. Eng.*, vol. 169, pp. 126–138, 2018.
31. M. U. H. Joardder, R. Alsbua, W. Akram, and M. A. Karim, "Effect of sample rugged surface on energy consumption and quality of plant-based food materials in convective drying," *Dry. Technol.*, pp. 1–10, 2020. doi:10.1080/07373937.2020.1745824
32. M. U. H. Joardder, C. Kumar, and M. A. Karim, "Prediction of porosity of food materials during drying: Current challenges and directions," *Crit. Rev. Food Sci. Nutr.*, vol. 58, no. 17, pp. 2896–2907, 2018.
33. J. Farkas, "Physical methods of food preservation," in *Food Microbiology: Fundamentals and Frontiers*, American Society of Microbiology, Third Edition, 2007, pp. 685–712.
34. J. M. Irudayaraj and S. Jun, *Food Processing Operations Modeling: Design and Analysis.* CRC Press, 2008.
35. M. Zhang, H. Jiang, and R.-X. Lim, "Recent developments in microwave-assisted drying of vegetables, fruits, and aquatic products—Drying kinetics and quality considerations," *Dry. Technol.*, vol. 28, no. 11, pp. 1307–1316, 2010.
36. N. Duc Pham, M. I. H. Khan, M. U. H. Joardder, A. M. N. Abesinghe, and M. A. Karim, "Quality of plant-based food materials and its prediction during intermittent drying," *Crit. Rev. Food Sci. Nutr.*, vol. 59, no. 8, pp. 1197–1211, 2019.
37. J. A. Hernández, "Optimum operating conditions for heat and mass transfer in foodstuffs drying by means of neural network inverse," *Food Control*, vol. 20, no. 4, pp. 435–438, 2009.
38. P. Zhao, S. Ge, D. Ma, C. Areeprasert, and K. Yoshikawa, "Effect of hydrothermal pretreatment on convective drying characteristics of paper sludge," *ACS Sustain. Chem. Eng.*, vol. 2, no. 4, pp. 665–671, 2014.
39. P. Perre and R. B. Keey, "Drying of wood: Principles and practices," in *Handbook of Industrial Drying*, CRC Press, 2006, pp. 846–903.

40. S. M. Demarchi, R. M. T. Irigoyen, and S. A. Giner, "Vacuum drying of rosehip leathers: Modelling of coupled moisture content and temperature curves as a function of time with simultaneous time-varying ascorbic acid retention," *J. Food Eng.*, vol. 233, pp. 9–16, 2018.
41. Y. Avramenko and A. Kraslawski, "Selection of process scheme and conditions for drying of pharmaceutical materials," in *Computer Aided Chemical Engineering*, vol. 26, Elsevier, 2009, pp. 165–170.
42. D. Rondot, "Heat and mass transfer in leather drying process," *Iran. J. Chem. Chem. Eng.*, vol. 23, no. 1, pp. 25–34, 2004.
43. T. Hotta, K. Nakahira, M. Naito, N. Shinohara, M. Okumiya, and K. Uematsu, "Origin of the strength change of silicon nitride ceramics with the alteration of spray drying conditions," *J. Eur. Ceram. Soc.*, vol. 21, no. 5, pp. 603–610, 2001.
44. H. N. Suresh, P. A. A. Narayana, and K. N. Seetharamu, "Conjugate mixed convection heat and mass transfer in brick drying," *Heat Mass Transf.*, vol. 37, no. 2–3, pp. 205–213, 2001.
45. K. Khuenpet, N. Charoenjarasrerk, S. Jaijit, S. Arayapoonpong, and W. Jittanit, "Investigation of suitable spray drying conditions for sugarcane juice powder production with an energy consumption study," *Agric. Nat. Resour.*, vol. 50, no. 2, pp. 139–145, 2016.
46. A. Langford, B. Bhatnagar, R. Walters, S. Tchessalov, and S. Ohtake, "Drying technologies for biopharmaceutical applications: Recent developments and future direction," *Dry. Technol.*, vol. 36, no. 6, pp. 677–684, 2018.
47. S. Pusat and H. H. Erdem, "Drying characteristics of coarse low-rank-coal particles in a fixed-bed dryer," *Int. J. Coal Prep. Util.*, vol. 37, no. 6, pp. 303–313, 2017.

2 The Physics in Drying

2.1 INTRODUCTION

Drying is the process of removing water from wet materials. It can be accomplished by providing a different form of energy, such as thermal, electromagnetic, and sound wave. During drying, energy is applied from different sources to remove water from wet materials. This incident energy increases the kinetic energy of water molecule towards the activation energy for phase change. This phenomenon of phase change is often known as evaporation. It is also classified as simultaneous heat and mass transfer phenomena.

During drying, energy can be transferred in three ways, as shown in Figure 2.1:

1. Energy transfer in the form of work
2. Energy transfer in the form of mass
3. Energy transfer in the form of heat

In these transport phenomena, many similarities exist among heat, mass, and momentum transfer. It is commonly observed that where there is an imbalance in a species, there is a transport phenomenon that takes place to help it reach equilibrium. This hidden tendency of equilibrium is often referred to as "driving force." In this regard, temperature, concentration, and velocity gradients are the driving forces of heat, mass, and momentum transfer, respectively.

$$\text{Flux} \propto \text{Driving force} \tag{2.1}$$

The transport phenomena follow the laws of conservation, including mass, energy, and momentum. All of the three modes of heat transfer, namely, conduction,

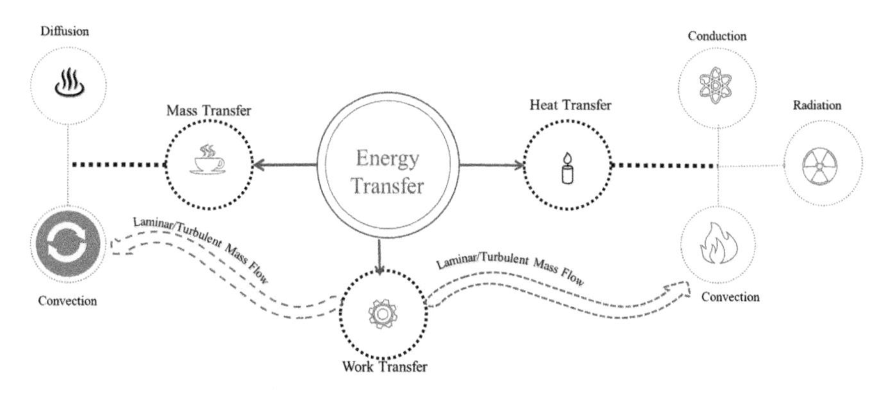

FIGURE 2.1 Different forms of energy transfer.

DOI: 10.1201/9780429461040-2

TABLE 2.1
Basic Transport Mechanisms and Driving Forces

Transport of	Driving Force	Flux Equation	Common Name
Mass (kg)	Concentration gradient	$J_m = -D \cdot \frac{dc}{dx}$	Fick's law of diffusion
Energy/heat (J)	Temperature gradient	$J_h = -k \cdot \frac{dT}{dx}$	Fourier's law of heat conduction
Momentum (kg·m/s)	Velocity gradient (shear rate)	$J_v = -\mu \cdot \frac{dv}{dx}$	Newton's law of viscosity
Volume (m^3)	Pressure gradient	$J = -L_p \cdot \frac{dP}{dx}$	Darcy's law

convection, and radiation, are involved in the drying process to supply heat energy to the sample. On the other hand, water removal from the core of the sample to the surface helps maintain different internal moisture transport phenomena. Fluid flow, especially of air, is vital when convection is key in transporting heat and mass. Table 2.1 shows the basic transport phenomena with their driving forces and common names.

The underlying physics that is essential for drying is briefly discussed in the following sections and described these in more detail where relevant in other chapters of this book.

2.2 MASS TRANSFER

Mass transfer refers to the movement of species due to a concentration gradient and is completely distinct from mass flow (fluid flow). Common unit processing operations including drying, crystallization, humidification, distillation, leaching, evaporation, and absorption, among others, are examples of mass transfer phenomena. Some of these processes, such as drying, involve heat transfer with mass transfer and are generally referred to as simultaneous heat and mass transfer phenomena.

There are a wide variety of mass transfer phenomena encountered during drying, and it is convenient to classify them based on their characteristics to make it reasonable to study them in groups. There are many ways to classify the internal mass transfer phenomena, and below we present some general types. These mechanisms have to be considered in different elemental arrangements in moisture-rich materials during drying. In many cases, several mass transfer mechanisms can account for moisture transfer as drying proceeds, though only one mechanism dominates moisture transfer at any given time.

Several mechanisms of internal mass transfer have been proposed in the drying literature, including liquid and vapour diffusion. Table 2.2 lists the internal mass transfer mechanisms that has been proposed for vapour and liquid phase

TABLE 2.2
Internal Moisture Transfer Mechanism during Drying of Solid Foods [1]

Reference	Moisture Transfer Mechanism during Drying	
Lewis (1921) [2]; Sherwood (1929) [3]	Diffusion	
Ceaglske and Hougen (1937) [4]	Capillary	
Henry (1948) [5]	Evaporation–condensation	
Görling (1958) [6]	*Transfer of liquid water* Capillary Liquid diffusion Surface diffusion	*Transfer of water vapour* Differences in partial pressure (diffusion) Differences in total pressure (hydraulic flow)
Van Arsdel (1963) [7]	Capillary Diffusion Surface diffusion in liquid layers absorbed at solid interfaces Water vapour diffusion in air-filled pores Gravity flow Vaporization–condensation sequence.	
Keey (1970) [8]	Hydraulic flow Capillary flow Evaporation–condensation Vapour diffusion	
Bruin and Luyben (1980) [9]	Molecular diffusion Capillary flow Knudsen flow Hydrodynamic flow Surface diffusion	
Hallström (1990) [10]	*Transfer of liquid water* Capillary (saturated) Molecular diffusion (within solid) Surface diffusion (absorbed water) Liquid diffusion (in pores) Hydraulic flow (in pores)	*Transfer of water vapour* Diffusion (in pores): Knudsen, Ordinary, Stephan diffusion. Hydraulic flow (in pores) Evaporation–condensation
Waananen et al. (1993) [11]	*Transfer of liquid water* Capillary diffusion Surface diffusion Hydraulic flow	*Transfer of water vapour* Mutual diffusion Knudsen flow Diffusion Slip flow Hydrodynamic (bulk) flow Stephan diffusion Poiseuille flow Evaporation–condensation

water transport. A summary of the common internal moisture transport mechanisms is presented in Table 2.3.

TABLE 2.3
Mechanisms of Moisture Transfer in Solid Food Materials [12–22]

Moisture Transfer Mechanisms		Visual Representation	Remarks
Diffusion	Stefan diffusion	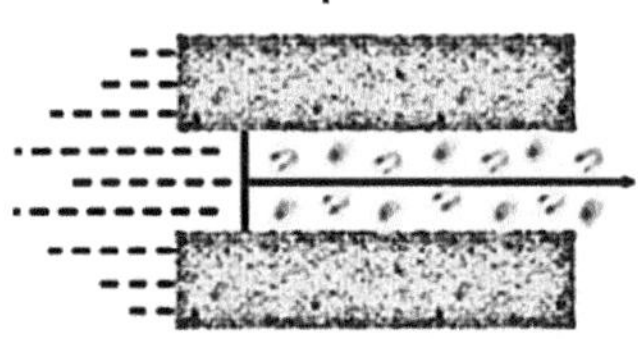	This diffusion phenomenon takes place in a multicomponent gaseous mixture [12]. The diffusion happens in a much more complex way than the one stated by Fick's law of diffusion [13].
	Mutual diffusion	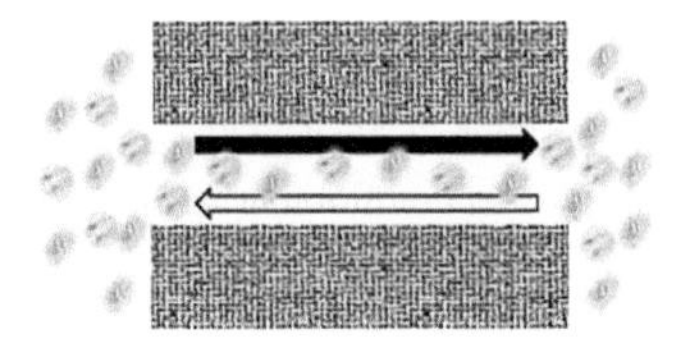	If a concentration gradient is established in a solution, it shows a tendency to reach equilibrium at a rate governed by a mutual diffusion. The mutual diffusion coefficient is related to the translation friction coefficient [14].
	Surface diffusion		This phenomenon of mass transfer takes place at the surface of a solid sample due to the concentration gradient [15]. Surface diffusion is an important phenomenon in heterogeneous catalysis [16,17].
	Knudsen diffusion	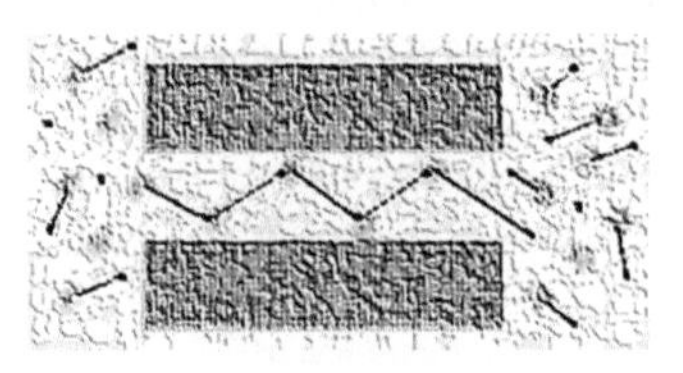	Knudsen diffusion is initiated when the pores of a porous solid are smaller than the mean free path of the gas or vapour molecules
	Liquid diffusion	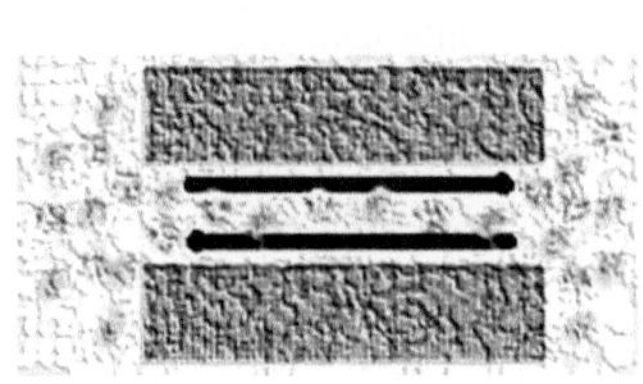	The diffusion mechanism in liquids is more complicated than that of its gaseous counterpart. This diffusion mechanism is related to the concentration gradient between liquids [18].

(*Continued*)

TABLE 2.3 (*Continued*)
Mechanisms of Moisture Transfer in Solid Food Materials [12–22]

Moisture Transfer Mechanisms		Visual Representation	Remarks
Convection	Poiseuille flow		Poiseuille flow is pressure-induced flow, often known as channel flow, in a long duct [19,20].
	Capillary flow	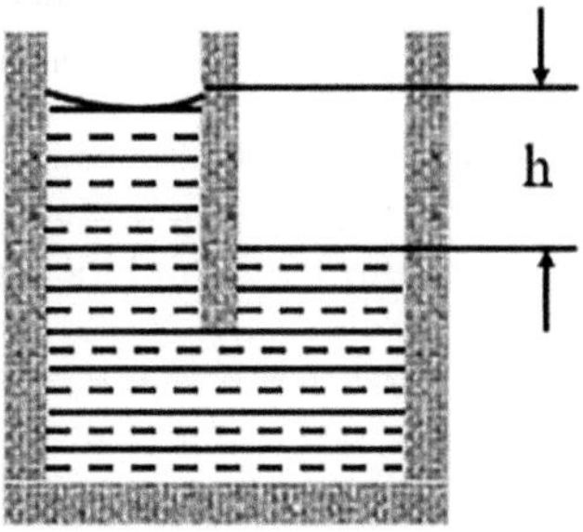 h	Capillary water movement is likely to be important at relatively high water contents that exceed saturation of liquid water. The flow of liquid into a capillary space is spontaneous and occurs without outside influence; rather, it depends on adhesion and cohesion forces [21]. When adhesion force is greater than cohesion force, capillary flow initiates.
Condensation evaporation		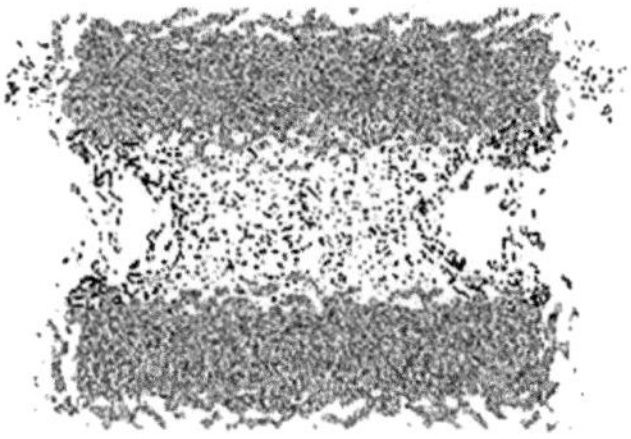	When a gas is cooled sufficiently or, in many cases, when the pressure on the gas is increased to saturation pressure, the gas condenses to the liquid phase. When a liquid is energized sufficiently or the pressure on the liquid is lower than saturation pressure, and the liquid evaporates to a gas [22].

There are many terms related to mass transfer and moisture that need to be defined for the convenience of the proceeding discussion. Moreover, the terms are confusing as several versions of the definition of a single term are available in different fields of drying. To maintain consistency in the meaning of the terms, a brief discussion is presented in the following section for common terminology related to mass transfer.

2.2.1 Mass Transfer-related Terminologies

2.2.1.1 Moisture Content

Moisture content is one of the most important factors that determines the extent of wetness of the sample. Generally, it can be expressed as the quantitative amount of water present in a sample on a wet or dry basis and can be expressed as follows:

$$\text{Wet basis} = \frac{\text{Weight}_{\text{wet}} - \text{Weight}_{\text{dry}}}{\text{Weight}_{\text{wet}}} \times 100 \tag{2.2}$$

$$\text{Dry basis} = \frac{\text{Weight}_{\text{wet}} - \text{Weight}_{\text{dry}}}{\text{Weight}_{\text{dry}}} \tag{2.3}$$

2.2.1.2 Water Concentration

The concentration of species can be determined from the mass of the species and the volume of the sample. Water concentration in a sample can be expressed in mole basis (kmol/m^3) or mass basis (kg/m^3). Either of the approaches can be taken as these are interchangeable.

2.2.1.3 Critical Moisture Content

Mass transfer rate varies with the variation of moisture content of the sample. At the beginning of drying, constant drying rate prevails, beyond which the drying rate begins to decline. The average moisture content from which point drying rate starts to decline is referred to as critical moisture content. In most cases, the critical moisture content is pretty much close to initial moisture content and thus does not hold significant importance. Different drying rate periods are shown in Figure 2.2 Reference?.

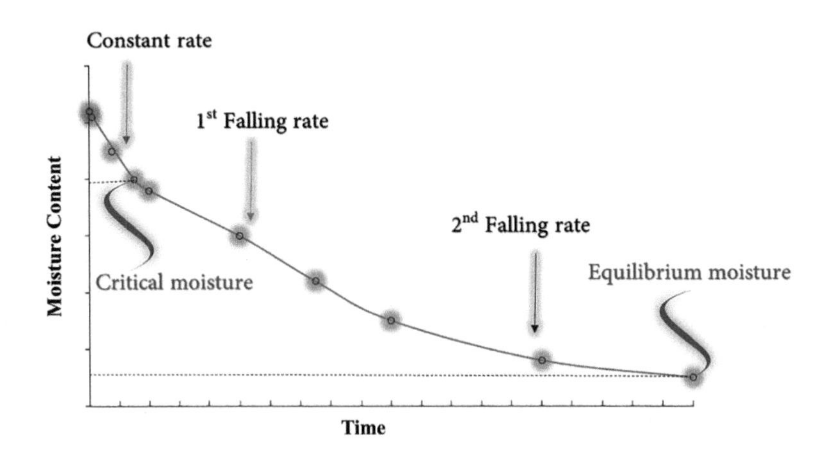

FIGURE 2.2 Different drying rate periods of a typical drying process.

The critical moisture content is strongly affected by the morphology of the sample, the proportion of bound and free water, sample thickness, and drying conditions, among other things.

2.2.1.4 Equilibrium Moisture Content

After the constant rate of drying, the falling rate starts. One or more falling rates can be observed during drying. With the progress of drying, moisture content reaches such an extent that there is no driving force of mass transfer in the sample. At this point, the vapour pressure of the sample is equal to that of the surrounding; moisture content of the sample at this juncture is referred to as equilibrium moisture content (EMC). Temperature, relative humidity, and water activity of the sample are the predominant factors determining equilibrium moisture content. In most cases, the EMC can be considered as the final moisture content of the dried material.

2.2.1.5 Moisture Sorption Isotherm

Moisture sorption isotherm (MSI) provides significant insight into thermodynamic behaviour, the specific energy requirement for drying, and water activity of the material. Moreover, the optimum drying condition can be determined from the MSI. From MSI, the nature of water interaction with the solid matrix can be demonstrated. Generally, a typical sorption isotherm (as shown in Figure 2.3) can be divided into three regions: bound water at region A, with region B treated as the transition zone from bound to free water, and loosely bound or free water molecules at region C.

2.2.1.6 Monolayer Moisture Content (MMC)

Water content can be categorized as monolayer moisture and multilayer moisture content, depending on the number of water layers on the solid matrix surface. Monolayer moisture content is sometimes regarded as the least amount of moisture suitable for dehydrated products. Most of the degradation phenomena of dried products are found negligible at or below MMC. In addition, removing MMC from products demands a significant amount of energy and causes several decomposition reactions. The presence of any additional layer of water along with the monolayer water is referred to as multilayer water. Removal of this/these additional layer(s) of water is much easier than mono-layer water removal.

2.2.1.7 Phases of Water

Water can be present in one or more of the three common forms of material, namely, solid, liquid, and vapour. In most cases, water is present as a mixture with other components and the solid matrix of the product determines its chemical behaviour. The purest form of water can only be achieved in the crystalline or vapour phase. Phase change needs a specific amount of energy equal to either the latent heat of vaporization or latent heat of sublimation, depending on the nature of the phase change. All liquid tends to convert to the gaseous phase, and all gas tends to condense into its liquid phase.

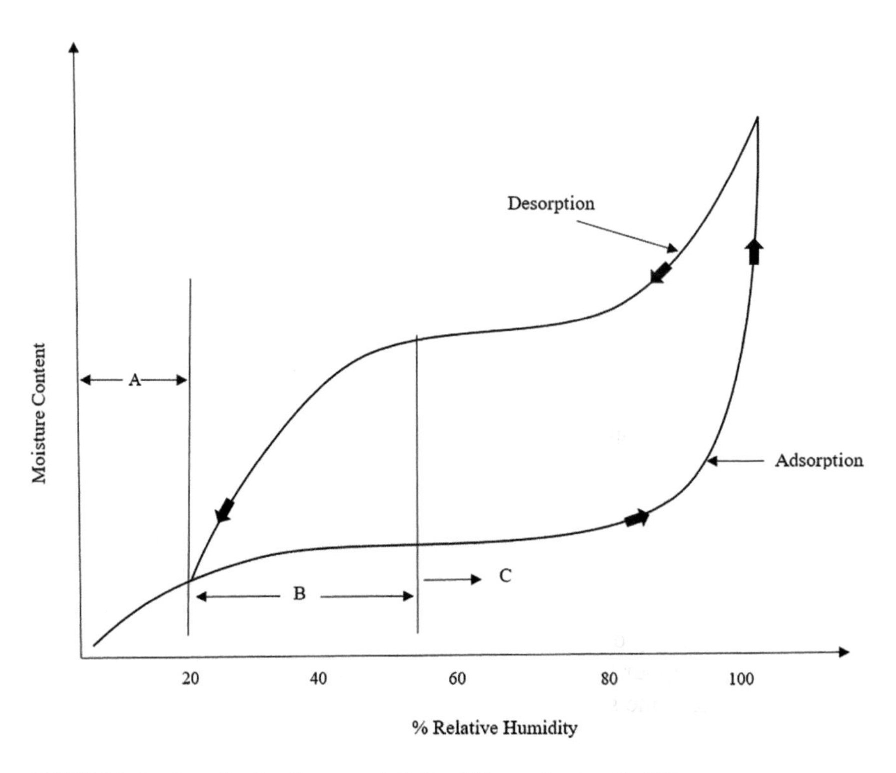

FIGURE 2.3 Sorption isotherm models for different foodstuffs [23].

Once the molecular kinetic energy of liquid molecules at the surface reaches its activation energy level, the phase starts changing from liquid to gas. This phenomenon is known as evaporation and the pressure caused by these vapour molecules is referred to as vapour pressure. Phase change phenomena can be categorized as evaporation and condensation, depending on the gradient of vapour pressure on the surface, as shown in Figure 2.4.

2.2.1.8 Water Potential

The concept of water potential considers all types of driving forces that cause movement of water from one place to another. Both macro- and micro-level driving forces, including hydrostatic pressure, capillary action, diffusion, and gravity, acting in the same or opposite directions, are taken into account in the water potential concept. In many products, during drying, diverse potentials contribute simultaneously. Solid matrix, composition, and atmospheric conditions significantly affect total water potential. Water potential of pure water is zero, and it decreases with the addition of solute, attaining a negative value. Total water potential (Ψ) can be expressed as a combination of capillary potential (Ψ_m), diffusion (Ψ_s), hydrostatic pressure (Ψ_p), and gravitational potential (Ψ_g).

$$\Psi = \Psi_m + \Psi_s + \Psi_p + \Psi_g \tag{2.4}$$

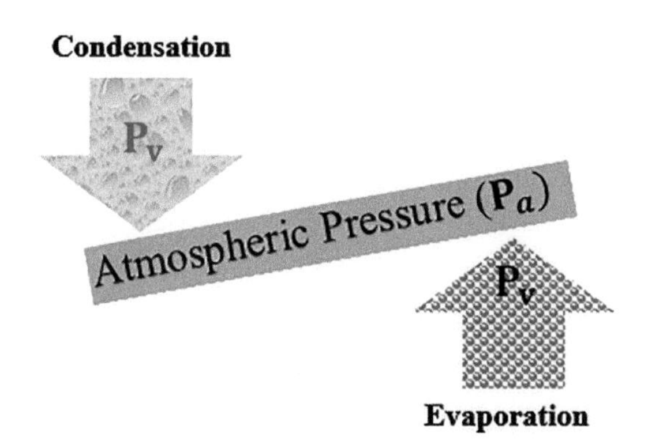

FIGURE 2.4 Roles of vapour pressure in evaporation and condensation.

2.2.1.9 Water Activity

The pattern of water distribution affects the stability of moist materials in terms of the response of microorganism growth and other physiochemical changes. It represents the overall tendency of migration of water from a sample. The water activity of a system is denoted a_w, and it can be determined from the ratio of the vapour pressure of water in sample to the vapour pressure of pure water at the same surrounding conditions. Following is the expression of water activity that correlate the vapour pressure of water in the sample with that of pure water.

$$a_w = \frac{P}{P_o} \tag{2.5}$$

Although the value of water activity refers to the information of the stability of the moist sample, it does not provide significant insight into the binding nature of water with another constituent of the system.

2.2.2 Types of Water

There are different types of water, and these are classified depending on the part of the sample where they exist, degree of mobility as well as bonding nature. Taking different classifications of water in foods into consideration, Joardder et al. [1] presented different proportions of water in a pyramid-like distribution, as demonstrated in Figure 2.5.

2.2.2.1 Free Water

Free water is the type of water that can easily be available in response to thermo-chemical processes, including drying. Moreover, free water accelerates the internal transport of other compounds and microbial growth. Less energy is required to remove free water from the sample. Free water is sometimes also named as bulk water and capillary water depending on the nature of the material.

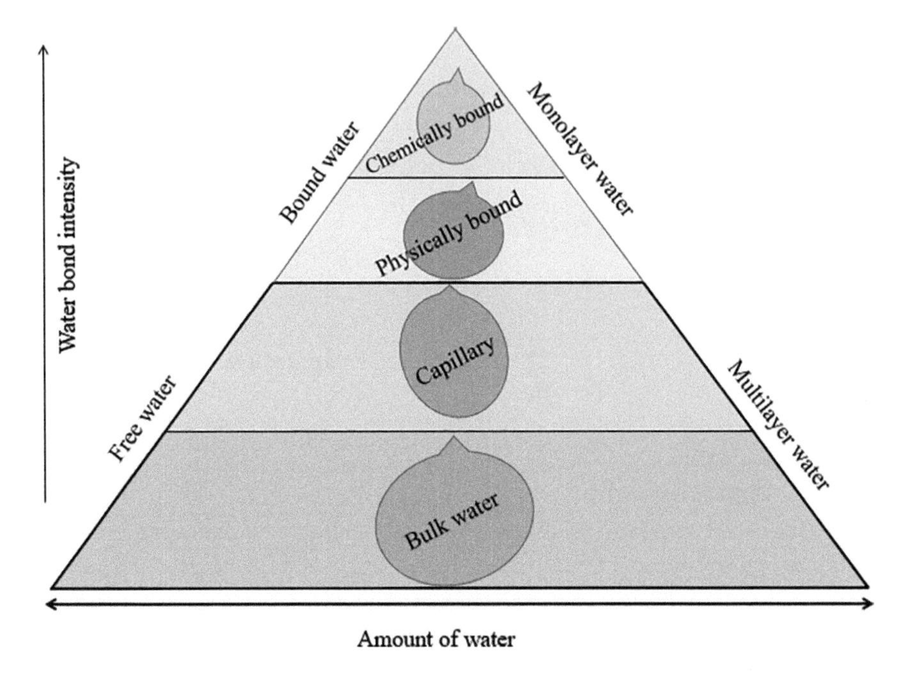

FIGURE 2.5 Different proportions of water in a pyramid-like distribution in food materials [1].

2.2.2.2 Bound Water

Defining bound water is not an easy task as there are several ways to define bound water in a moist sample. Most definitions of bound water refer to the water that is closely attached to other compounds of samples and shows distinguishing properties from the remaining "free water". The bound water is further classified as *strongly bound* and *loosely bound water* depending on the location, as shown in Figure 2.5. Bound water can also be classified as being physically bound and chemically bound. The interested readers are referred to advanced books on the topic for detailed discussions [24].

2.2.2.3 Spatial Distribution of Water

Water can be categorized based on the spatial location of their existence in the moist sample. For example, in biological tissue, water can be present at the cell wall and intracellular and intercellular spaces, as shown in Figure 2.6. The characteristics of water such as freezing point, specific energy consumption to evaporate, and susceptibility to microorganisms differ based on variation of space.

It is worth mentioning that the spatial distribution of water varies with the variation of material. The water distribution in non-hygroscopic materials varies from the water distribution in hygroscopic materials.

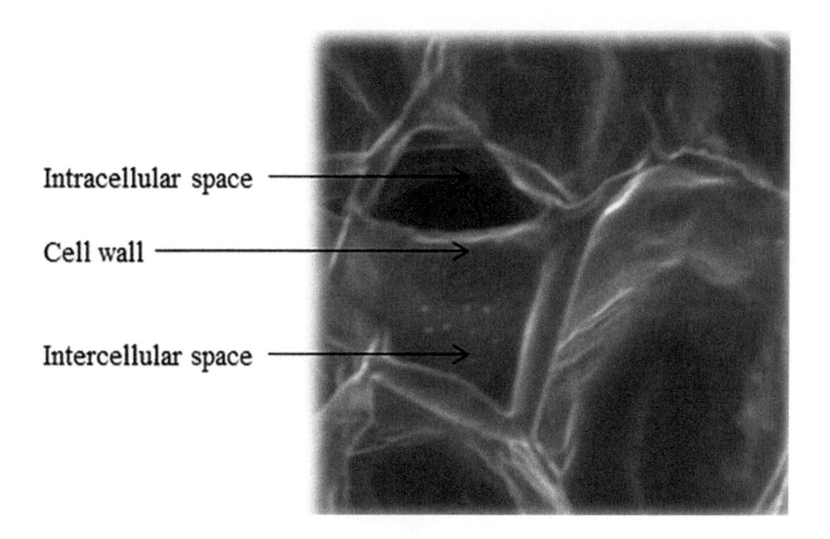

FIGURE 2.6 Spatial distribution of water in cellular foods [1].

2.3 HEAT TRANSFER PHENOMENA DURING DRYING

Heat is applied for a wide range of purposes including moisture removal, reducing enzymatic activity, increasing bioavailability, and increasing the shelf-life of consumable materials, including foods. Although heating and cooling serve different objectives in food processing, they share the same form of heat transfer, as shown in Figure 2.7. In most drying operations, one or more modes of heat transfer phenomena take place to drive away moisture from wet objects. In the following section, we will provide a brief description of each mode to acquaint the reader with the basic mechanisms of heat transfer. The reader interested in more insight into the modes of heat transfer can consult numerous other sources available, including heat and mass transfer textbooks.

Heat transfer is a very common phenomenon that happens between two regions that temperature differences. Therefore, the temperature gradient is the driving force of heat transfer phenomena. In general, heat travels from higher temperatures to lower temperatures. However, in radiation heat transfer, heat can travel in both directions due to having a greater than absolute zero temperature.

Convection heat transfer is a surface phenomenon where fluid flow is involved in moving heat from the surface of an object to the surrounding. Convection heat transfer significantly depends on the types of fluids, the velocity of fluid, and types of surfaces, among other things. Convection heat transfer is classified as natural and forced, depending on whether the fluid is forced to flow or is the result of buoyancy forces due to the effect of temperature.

On the other hand, thermal radiation is an exceptional heat transfer method that does not require any medium, unlike conduction and convection. Any object

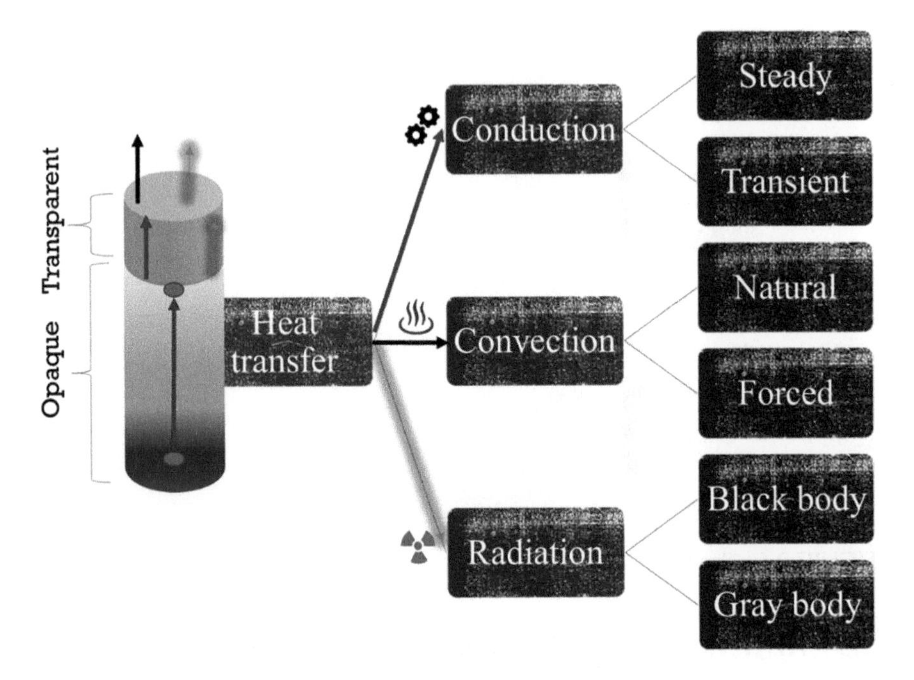

FIGURE 2.7 Different modes of heat transfer.

with a temperature of more than absolute zero spontaneously emits thermal energy in the form of electromagnetic waves. The direction of thermal radiation does not depend on the temperature of the objects as is the case in convection and conduction heat transfer. It is the fastest mode of heat transfer as it does not need any medium and has high speed electromagnetic waves.

2.3.1 Conduction Heat Transfer

Conduction happens in both solid and fluid provided there is no bulk motion prevalent in the fluid. The temperature difference in solids causes vibration of lattice and flow of free electrons, resulting in heat conduction from a higher temperature zone to a lower temperature zone. On the other hand, collision and diffusion of molecules in fluid result in conduction of heat in the fluid medium. During drying, a cold sample eventually warms up to the drying air temperature as a result of convection and conduction. The surface temperature increases due to convection and the propagation of heat towards the centre of the sample, caused by conduction.

The rate of heat conduction in a medium mainly depends on the shape and size of the sample, thermal properties, and temperature difference across the sample. Based on spatial and temporal characteristics, conduction heat transfer can be classified as steady, lump system, and transient, as shown in Figure 2.8.

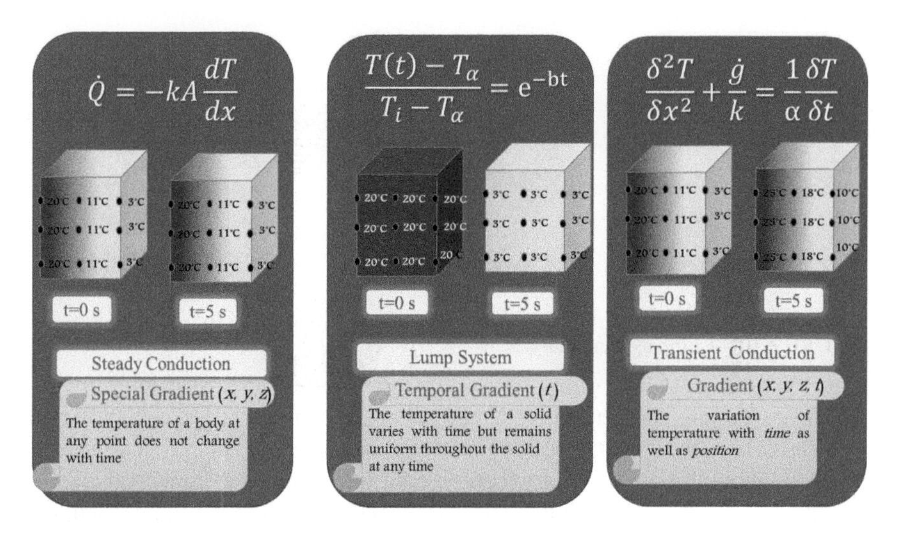

FIGURE 2.8 Different forms of conduction heat transfer.

2.3.1.1 Steady Conduction

When the temperature of a system varies spatially and remain constant at any point with time it is categorized as steady conduction. The conductivity of materials is one of the most important factors that determines how rapidly a substance conducts heat through the system . For this obvious reason, the heat transfer rate in a solid and static fluid is directly proportional to its conductivity. The heat transfer rate is also proportional to the temperature gradient across the system and the area (A) normal to the direction of heat transfer. Taking all these considerations, Fourier's law of heat conduction is expressed in Figure 2.9.

As heat is transferred from increasing temperature, the gradient of temperature is negative while heat is conducted in a positive direction. Although the conductivity varies with temperature, an average conductivity can be used for a system with a low-temperature gradient. Similarly, structural heterogeneity also affects conductivity throughout the sample. Therefore, for isotopic materials, such as most engineering materials, conductivity remains constant throughout

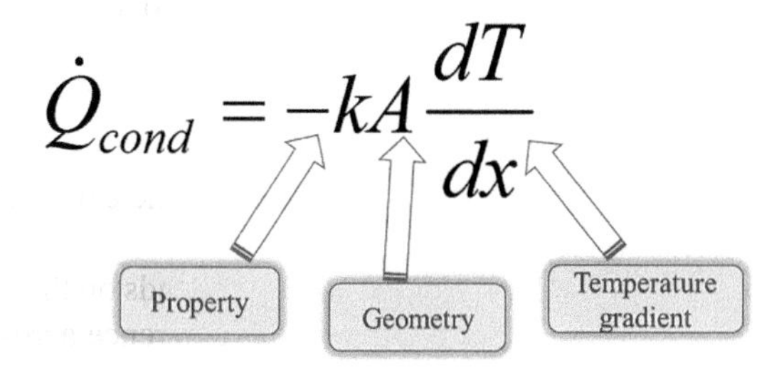

FIGURE 2.9 Fourier's law of heat conduction.

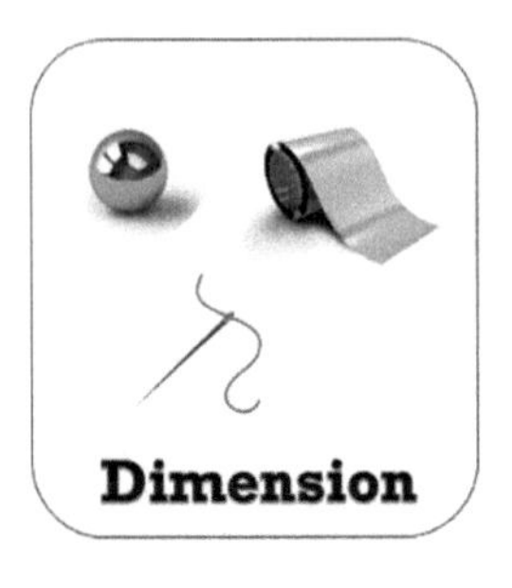

FIGURE 2.10 Lump system analysis.

the entire sample. On the other hand, anisotropic materials like plant-based materials show the variation of conductivity and other thermophysical properties throughout the sample. In such cases, consideration of constant conductivity would provide erroneous result. Variable thermal conductivity will be discussed in Chapter 3.

2.3.1.2 Lump System

Unlike the steady system, the temperature remains constant across the sample in a lumped system. However, temperature varies with time in lump systems analysis. Lump system is applicable for a sample with high conductivity, smaller sample, and natural convection boundary condition as shown in Figure 2.10.

As most water-rich samples show low conductivity for force convection, lump system analysis should not be applied to determine the temperature distribution during drying.

2.3.1.3 Transient Heat Conduction

Transient heat conduction considers both spatial and temporal change of temperature in the sample during drying. This is the most realistic approach to conduction heat transfer. The governing equation for transient heat conduction is presented in Table 2.4.

What causes the variation in governing equation with changes of shapes? It is due to the variation of the area in the direction of heat conduction. To clarify this, we can take an onion, a roll of tissue, and a book as corresponding to a sphere, a cylinder, and a plane wall, as shown in Figure 2.11.

Peeling the onion and unrolling the tissue result in the exposed surface being increased; eventually, there should be a gradient of the area in the governing equation as shown in Table 2.4. However, turning pages does not cause a decrease in the normal area, and there is no gradient of area present in the governing equation, as shown in the following expression:

$$\rho C_p\left(\frac{\partial T}{\partial t}+\boldsymbol{u}.\Delta T\right)=\Delta.(k\Delta T)+Q \tag{2.6}$$

TABLE 2.4
Basic Shapes with Their Governing Equation [25]

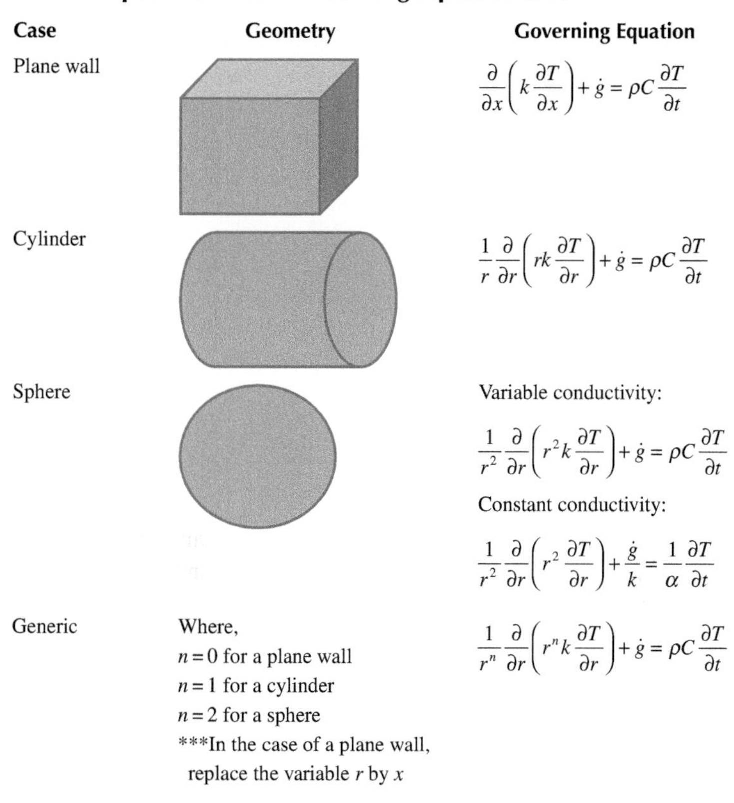

Case	Geometry	Governing Equation
Plane wall		$\frac{\partial}{\partial x}\left(k\frac{\partial T}{\partial x}\right)+\dot{g}=\rho C\frac{\partial T}{\partial t}$
Cylinder		$\frac{1}{r}\frac{\partial}{\partial r}\left(rk\frac{\partial T}{\partial r}\right)+\dot{g}=\rho C\frac{\partial T}{\partial t}$
Sphere		Variable conductivity: $\frac{1}{r^2}\frac{\partial}{\partial r}\left(r^2k\frac{\partial T}{\partial r}\right)+\dot{g}=\rho C\frac{\partial T}{\partial t}$ Constant conductivity: $\frac{1}{r^2}\frac{\partial}{\partial r}\left(r^2\frac{\partial T}{\partial r}\right)+\frac{\dot{g}}{k}=\frac{1}{\alpha}\frac{\partial T}{\partial t}$
Generic	Where, $n=0$ for a plane wall $n=1$ for a cylinder $n=2$ for a sphere ***In the case of a plane wall, replace the variable r by x	$\frac{1}{r^n}\frac{\partial}{\partial r}\left(r^nk\frac{\partial T}{\partial r}\right)+\dot{g}=\rho C\frac{\partial T}{\partial t}$

FIGURE 2.11 Onion, tissue roll, and book corresponding to the sphere, cylinder, and plane wall.

If the velocity is set to zero, the equation governing purely conductive heat transfer is obtained:

$$\rho C_p\frac{\partial T}{\partial t}+\Delta.(-k\Delta T)=Q \quad (2.7)$$

2.3.2 Convection Heat Transfer

Generally, convection heat transfer is a surface phenomenon. When heat transfers from a surface is collectively caused due to heat diffusion as well as bulk fluid motion, then it is referred to as convection heat transfer. It is the simultaneous heat and momentum transfer phenomena. In the absence of bulk motion of the fluid, the heat transfer between the adjacent fluid and solid surface can be treated as pure conduction. Eventually, both conduction and convection heat transfer phenomena can take place in fluid media depending on the nature of fluid motion. Therefore, the heat transfer rate depends on fluid type, pattern of fluid flow, and surface condition.

As convection heat transfer is a surface phenomenon, thermal properties of solid materials do not contribute to the heat transfer rate coefficient. Depending on the way of fluid motion initiation, convection is classified as free or forced convection. In free convection, the fluid motion is caused without any external means, while forcing convection is initiated by external means such as a fan.

Determining the convection heat transfer rate is quite difficult as it depends on several of the mentioned variables. It can be determined using the dimensionless numbers such as Reynolds, Prandtl, and Nusselt numbers. These equations carry the effect of convection heat transfer-associated parameters including type of fluid, flow pattern, as well as surface condition.

From the given conditions, Reynolds and Prandtl numbers can be used to obtain an appropriate empirical relationship with a Nusselt number. Following this, the convection heat transfer coefficient can be calculated from steps 3 and 4 as shown in Figure 2.12. The interested reader is referred to heat transfer textbooks on the topic for detailed discussions.

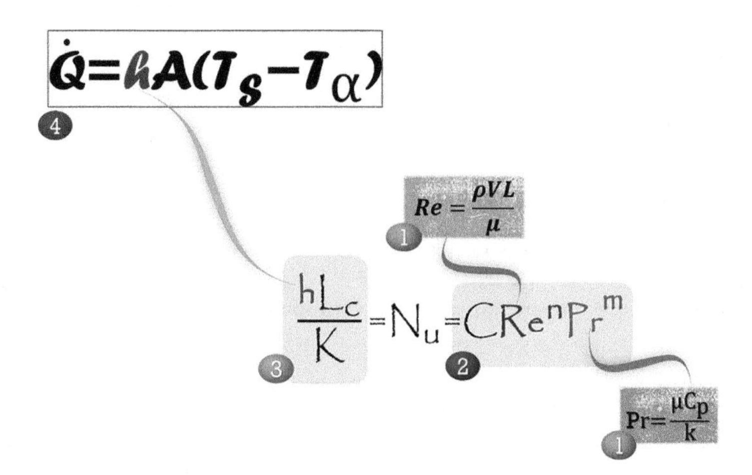

FIGURE 2.12 Steps associated with determining the convection mass transfer rate.

For example, the heat transfer coefficient can be calculated from the following relationships for laminar (Eq. 2.8) and turbulent (Eq. 2.9) flows over a flat plate, respectively:

$$\mathrm{Nu} = \frac{hL}{k} = 0.332\,\mathrm{Re}^{0.5}\,\mathrm{Pr}^{0.33} \tag{2.8}$$

$$\mathrm{Nu} = \frac{hL}{k} = 0.0296\,\mathrm{Re}^{0.5}\,\mathrm{Pr}^{0.33} \tag{2.9}$$

For the Nusselt number for cross-flow over a cylinder, we present the equation proposed by Churchill and Bernstein [25]:

$$\mathrm{Nu}_{\mathrm{cyl}} = \frac{hD}{k} = 0.3 + \frac{0.62\,\mathrm{Re}^{\frac{1}{2}}\,\mathrm{Pr}^{\frac{1}{3}}}{\left[1+\left(\frac{0.4}{\mathrm{Pr}}\right)^{\frac{2}{3}}\right]^{\frac{1}{4}}}\left[1+\left(\frac{\mathrm{Re}}{2,82,000}\right)^{\frac{5}{8}}\right]^{\frac{4}{5}} \tag{2.10}$$

This relation is quite comprehensive in that it correlates available data well for Re Pr > 0.2. The fluid properties are evaluated at the *film temperature* $T_f = \frac{1}{2}(T_\infty + T_s)$, which is the average of the free-stream and surface temperatures.

For flow over a *sphere*, Whitaker recommends the following comprehensive correlation [25]:

$$\mathrm{Nu}_{\mathrm{sph}} = \frac{hD}{k} = 2 + \left[0.4\,\mathrm{Re}^{\frac{1}{2}} + 0.06\,\mathrm{Re}^{\frac{2}{3}}\right]\mathrm{Pr}^{0.4}\left(\frac{\mu_\infty}{\mu_s}\right)^{\frac{1}{4}} \tag{2.11}$$

which is valid for $3.5 \leq \mathrm{Re} \leq 80{,}000$ and $0.7 \leq \mathrm{Pr} \leq 380$. The fluid properties, in this case, are evaluated at the free-stream temperature T_∞, except for μ_s, which is evaluated at the surface temperature T_s.

2.3.3 Radiation Heat Transfer

Radiation heat transfer is associated with many drying systems directly or indirectly. Sun drying, solar drying, and infrared drying are common drying methods that are significantly associated with radiation heat transfer phenomena. In the case of opaque materials, such as most food materials, radiation is considered to be a surface phenomenon.

Irradiated energy on such surfaces and emitted energy by the interior regions of such material cannot cross the surface, except for being absorbed to within a few microns from the surface. Unlike the other two modes of heat transfer, the presence of a *medium* such as a solid or fluid substance is not essential in transferring energy by radiation. A body that is at a temperature of over 0 K (absolute zero temperature) can emit thermal radiation.

Thermal radiation encompasses the portion of the electromagnetic spectrum ranging from about 0.1 to 100 μm. Of this, infrared heating utilizes the wavelength range of 0.7–100 μm. Infrared heating is widely used in the food industry for drying of many foodstuffs, especially, vegetables, fruits, fish fillets, and grain. Far-infrared radiation (FIR) is used to dry the foods in the form of thin layers such as leafy vegetables, whereas near-infrared radiation (NIR) is used to dry thicker-bodied fruits.

At a given wavelength, the amount of energy radiated from a surface depends on the surface temperature and conditions of the surface. There is no surface that can emit more energy than a blackbody at a specified temperature and wavelength. Regardless of the wavelength and direction, a blackbody can absorb all incident radiation. Radiation from a blackbody can be determined from the following equation [25]:

$$E_b(T) = \sigma T^4 \ (\mathrm{W/m^2}) \tag{2.12}$$

The above relation is known as the Stefan-Boltzmann law, which refers to the emission of thermal radiation as being proportional to the fourth power of the absolute temperature. If the surface temperature is T_s (K), the maximum rate of radiation that can be emitted from a surface can be determined by the following equation:

$$\dot{Q}_{emit,\max} = \sigma A_s T_s^4 \ (\mathrm{W}) \tag{2.13}$$

At the same temperature, the real surface radiation is always less than the blackbody radiation which can be expressed as follows [26]:

$$\dot{Q}_{emit} = \varepsilon \sigma A_s T_s^4 \ (\mathrm{W}) \tag{2.14}$$

where ε is the emissivity of the surface. The values of emissivity of different materials range from 0 to 1 and are used to measure how closely a surface approximates a blackbody. Table 2.5 shows the emissivity values of different materials commonly used in drying.

If a sample is enclosed by a much larger surface at temperature T_{surr}, the net rate of radiation heat transfer between these two surfaces is given by the following equation (Figure 2.13):

$$\dot{Q}_{rad} = \varepsilon \sigma A_s \left(T_s^4 - T_{surr}^4\right) \ (\mathrm{W}) \tag{2.15}$$

The above equation can be applied in infrared and similar types of drying. However, for more accurate hear transfer from the realistic body, radiation from the grey body is the accurate approach, as discussed in the following section.

TABLE 2.5
Emissivity of Different Drying Materials [25,26]

	Material	Emissivity
Food	Beef	0.73–0.74
	Frozen foods	0.95
	Food	0.85–1.00
Wood		0.80–0.95
Textile		0.85–0.98
Paper		0.92–0.97
Leather		0.95–1.00
Brick		0.90–0.96
Ceramics		0.90–0.95
Powder		0.96
Coal		0.80–0.95
Water		0.95–0.963

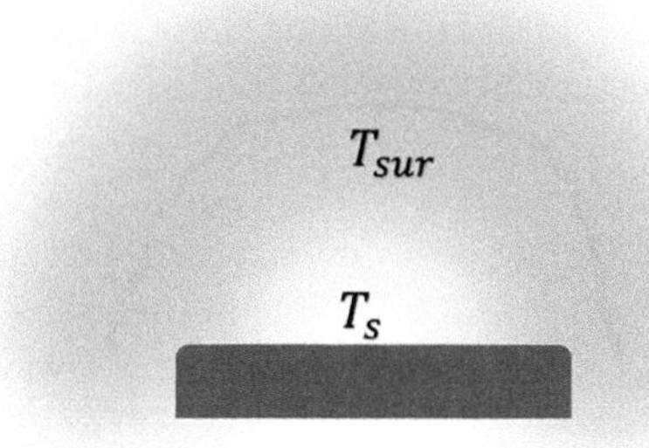

FIGURE 2.13 Radiation heat transfer between the surface and surrounding area.

2.3.3.1 Grey Body Heat Radiation

As seen in the previous section, the analysis of radiation transfer in enclosures consisting of black surfaces is relatively easy. On the other hand, in practice, most enclosures are nonblack surfaces that allow multiple reflections to occur. Unless some simplifying assumptions are made, the radiation analysis of such enclosures becomes very complicated. It is common to assume the surfaces of an enclosure to be opaque, diffuse, and grey because of making simple radiation analysis of the surface. Thus, the surfaces should be nontransparent, diffuse emitters, and diffuse reflectors, and their radiation properties independent of wavelength. Besides, each of the enclosure surfaces is isothermal, and both the incoming and outgoing radiation is uniform over each surface [25].

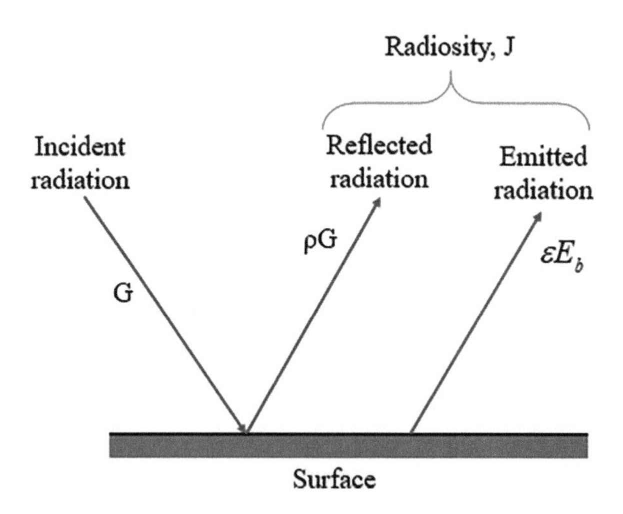

FIGURE 2.14 The sum of radiation energy reflected and emitted is radiosity.

Since the surfaces emit and reflect radiation, the radiation leaving a surface consists of emitted and reflected parts. The calculation of radiation heat transfer between surfaces involves the total radiation energy streaming away from a surface, with no regard to its origin. The total radiation energy leaving a surface per unit time and per unit area is known as radiosity and expressed by J (Figure 2.14) [25].

If the surface resistance of two bodies and the space resistance between them is considered, then the net heat flow can be represented by the circuit as shown in Figure 2.15. The net heat exchange between the two grey surfaces is represented by:

$$(Q_{12})_{net} = \frac{E_{b1} - E_{b2}}{\frac{1-\epsilon_1}{A_1 \epsilon_1} + \frac{1}{A_1 F_{12}} + \frac{1-\epsilon_2}{A_2 \epsilon_2}}$$

2.4 MASS TRANSFER BASICS

Nature is created in such a way that it has a constant tendency to achieve equilibrium. When there exists a concentration difference among species, the mass transfer starts from the higher concentration and proceeds to the lower concentration zone. Sometimes, the mass transfer is facilitated by the flow of other fluid, and thus its rate is increased. When mass transfer is caused due to a concentration difference, it is referred to as diffusion mass transfer. The mass transfer that is associated with a bulk fluid flow is categorized as convective mass transfer [27,28].

2.4.1 DIFFUSION

Mass transfer is mainly proportional to the *concentration gradient* of the species and is also affected by the surface area normal to the direction of mass transfer.

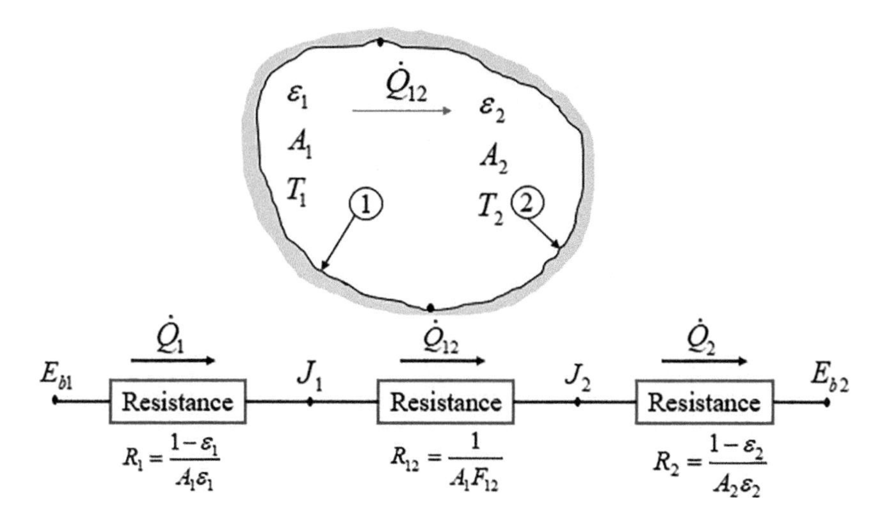

FIGURE 2.15 Schematic of a two-surface enclosure and the radiation network associated with it [25].

Generally, the diffusion mass transfer rate is calculated from the well-known Fick's laws of diffusion.

$$\dot{Q} = -k_{diff} A \frac{dC}{dx} \tag{2.16}$$

Here, the proportionality constant k_{diff} is the *diffusion coefficient* of the medium. This refers to how well a species can diffuse in a medium. Higher diffusivity denotes the faster speed of the species during mass transfer. There are two distinct ways of expressing the concentration of a species, namely, mass basis and molar basis, as shown in Table 2.6. Although both approaches are interchangeable and equivalent, the option that provides an easier solution to a given problem should be chosen.

In solid diffusion, a fluid or solute transfers from the homogeneous solid structure, for instance, leaching of soybean oil from soybean-solvent solution. The total flux is given by:

$$N_A = -D_{AB} \frac{dC_A}{dz} \tag{2.17}$$

where D_{AB} is the binary diffusivity of A in solid structure B. In many cases, the diffusivity of fluids is affected by pore characteristics including size and type. When the path of diffusion is not straight rather having tortuosity (τ), the diffusivity decreases significantly. If ε is the porosity of the solid matrix, the effective diffusivity can be expressed as follows:

$$D_{Ae} = \frac{\varepsilon}{\tau^2} D_{AB} \tag{2.18}$$

TABLE 2.6
Different Ways of Expressing the Concentration of a Species

Way of Expressing Concentration	Partial Concentration	Total Concentration	Fraction of Species	Notation
Mass	$\rho_i = \frac{m_i}{V}$ (kg/m³)	$\rho = \frac{m}{V} = \sum \frac{m_i}{V} = \sum \rho_i$	$w_i = \frac{m_i}{m} = \frac{\frac{m_i}{V}}{\frac{m}{V}} = \frac{\rho_i}{\rho}$	m = mass M = molar Mass N = mole number w = mass fraction
Molar	$C_i = \frac{N_i}{V}$ (kmol/m³)	$C = \frac{N}{V} = \sum \frac{N_i}{V} = \sum C_i$	$y_i = \frac{N_i}{N} = \frac{\frac{N_i}{V}}{\frac{N}{V}} = \frac{C_i}{C}$	y = mole fraction V = volume ρ = density C = concentration Subscript i = species
Interrelationship	$w_i = \frac{\rho_i}{\rho} = \frac{C_i M_i}{CM} = y_i \frac{M_i}{M}$			

2.4.1.1 Transient Mass Diffusion

Most diffusion that takes place is the transition in nature. For example, mass transfer in the drying of food, timber, coal, and textile is governed by transient mass transfer. In this case, the concentration of species not only varies with space but also varies with time. From Fick's second law, the following relationship can be established:

$$\frac{\partial C}{\partial t} = D_{eff} \frac{\partial^2 C}{\partial x^2} \tag{2.19}$$

In the presence of a bulk flow of the species, the above equation can be expressed as follows.

$$\frac{\partial M}{\partial t} + u\frac{\partial M}{\partial x} = D_{eff} \frac{\partial^2 M}{\partial x^2} \tag{2.20}$$

In the equation, the term with u contributes to additional convection mass transport mechanism. If the generation of species takes place due to chemical reaction or evaporation, an additional source term R needs to be incorporated, and we can write the above equation as follows:

$$\frac{\partial M}{\partial t} + \left(-D_{eff} \frac{\partial^2 M}{\partial x^2}\right) + u\frac{\partial M}{\partial x} = \pm R \tag{2.21}$$

The analytical solution of Equation (2.19) is almost impossible and time-consuming. In many situations of drying, when the drying of the surface and its adjacent layer is a concern, the sample can be treated as a semi-infinite medium. For a semi-infinite medium with constant surface concentration, the solution can be expressed from the steps that are shown in Figure 2.16.

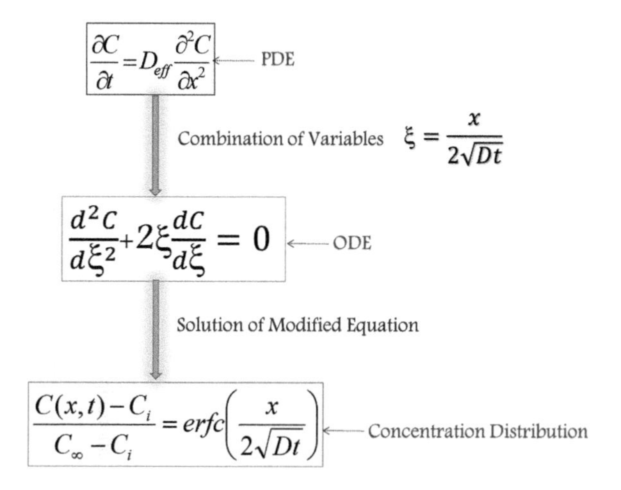

FIGURE 2.16 Concentration distribution determination in semi-infinite solid.

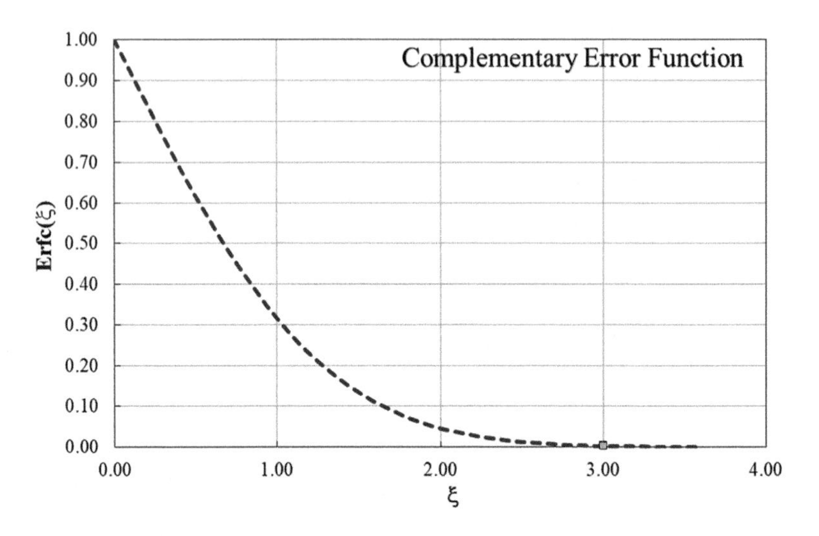

FIGURE 2.17 Complementary error function to find out the combined variable.

To figure out the value of a combined variable, the complementary error function is generally used as shown in Figure 2.17.

2.4.2 Mass Convection

When a mass transfer from a surface is collectively caused due to mass diffusion as well as bulk fluid motion, then it is referred to as convection mass transfer. Basically, it is the simultaneous mass and momentum transfer phenomena. Mass transfer rate in convection removes high-concentrated (species) air and provides low-concentrated one, resulting in high mass transfer. Figure 2.18 shows the internal and external mass transfer mechanism during drying.

To determine convection mass transfer, we need to use empirical equations as this type of mass transfer is associated with several factors such as shape and

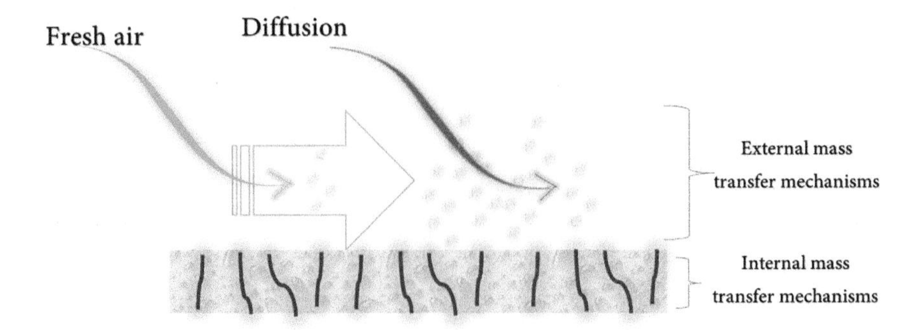

FIGURE 2.18 Internal and external mass transfer mechanism.

TABLE 2.7
Necessary Non-Dimensional Parameters to Analyze Heat and Mass Transfer-Related Problems [25]

Number	Symbol	Group	Typical Relevance	Parameters
Reynolds	Re	$\frac{\rho VX}{\mu}$	In all flows	Velocity of fluid, V Density of fluid, ρ
Nusselt	Nu	$\frac{hX}{k}$	Convection heat transfer	Viscosity of fluid, μ Thermal conductivity of fluid, k
Prandtl	Pr	$\frac{C_p\mu}{k}$	Fluid properties in determining heat transfer coefficient	Dimension of surface, X Heat transfer coefficient, h Specific heat capacity of fluid, C_p
Schmidt	Sc	$\frac{v}{D_{AB}}$	Fluid properties in determining mass transfer coefficient	Viscosity of fluid, μ Coefficient of expansion of fluid, β Force of gravity, g
Sherwood	Sh	$\frac{VL}{D_{AB}}$	Determining mass transfer coefficient	Temperature difference, θ Characteristic length, L, X Thermal diffusivity, α
Lewis Number	Le	$\frac{\alpha}{D_{AB}}$	Simultaneous heat and mass transfer	Kinematic viscosity of fluid, υ Mass diffusivity, D_{AB}

size of the sample, surface roughness, flow type, fluid type, and its properties, as shown in Table 2.7.

As the mechanisms of convection heat and mass transfer are analogous to each other, a clear understanding of convection mass transfer can be attained with a little effort. It is so analogous that one can even use the same equation for a certain condition by replacing the corresponding dimensionless number; for instance, Schmidt number is used instead of Prandtl number, as shown in Figure 2.19. In many cases, the concept and calculation of convection mass transfer can be attained by simply replacing the word 'heat' with 'mass'.

The heat and mass flow analogy is valid in most of the drying phenomena. Convection mass transfer mainly dominates the external mass flow rate. Apart from a concentration gradient, convective mass transfer coefficient plays a significant role in overall convective mass transfer.

2.5 FLUID FLOW

Species movement can be possible through both mass transfer and mass flow. The distinction is maintained due to their driving force, as shown in Figure 2.20. The driving force of fluid flow is the pressure difference, and it occurs at the macroscopic level. Water in wet materials sometimes flows through a narrow channel that is distinct from the mass transfer. During drying, fluid flow needs to be considered both in food materials and the surrounding of food materials. The drying

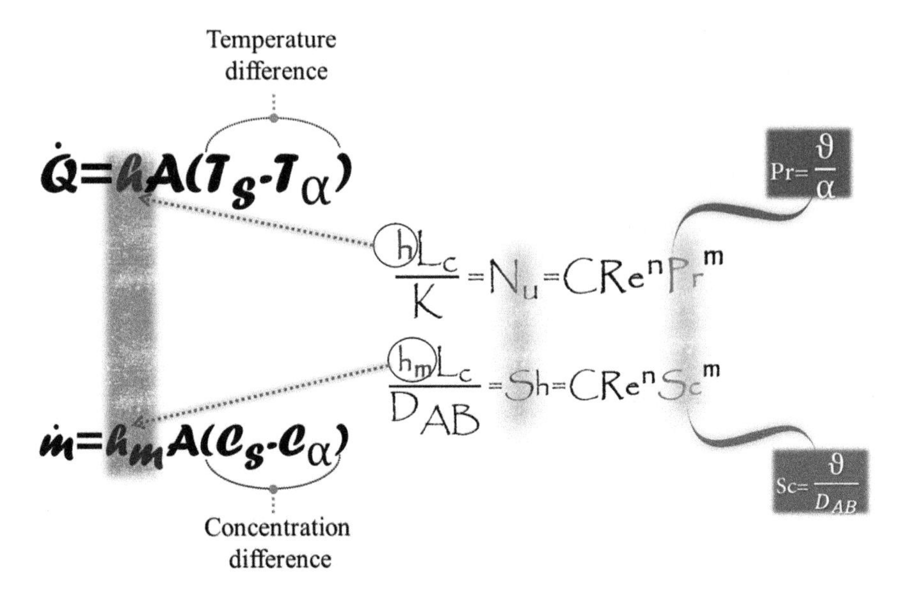

FIGURE 2.19 Analogy between heat and mass transfer.

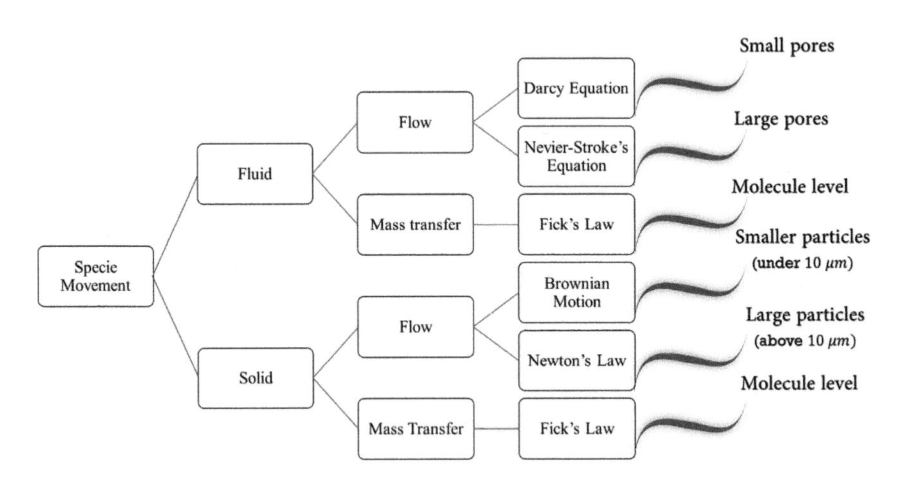

FIGURE 2.20 Schematic showing various ways of species movement.

system, which uses airflow for moisture removal and fluid flow modelling, needs to be simulated in a conjugate problem.

Darcy's law is mathematically parable with Fick's law of diffusion and Fourier's law of conduction. Equation (2.22) is known as Darcy's equation and is commonly used determining volumetric flow rate Q (m^3/s).

$$\dot{Q} = -k\ A \frac{\Delta(P - \rho g z)}{L} \tag{2.22}$$

TABLE 2.8
Comparison among Darcy's Law, Brinkman Equation, and Richards' Equation

Name of Equation		Equations	Remarks
Porous media flow	Darcy's law	$\frac{\partial}{\partial t}(\in_p \rho)+\nabla\cdot(\rho u)=Q_m$ $u=-\frac{\kappa}{\mu}\nabla p$	Low-velocity flow Low permeability Very small porosity Driving force: pressure gradient
	Brinkman equations	$\rho\frac{\partial u}{\partial t}=\nabla\cdot[-pl+K]$ $-\left(\mu\kappa^{-1}+\beta\in_p\rho\lvert u\rvert+\frac{Q_m}{\in_p^2}\right)$	Extension of Darcy's Equation Viscous shear is considered in kinetic energy dissipation Applicable for compressible fluids also Both velocity and pressure field are considered
	Richards' equation	$\frac{\partial}{\partial t}(\in_p \rho)+\nabla\cdot(\rho u)=Q_m$ $u=-\frac{\kappa}{\mu}(\nabla p+\rho g)$	Applicable for variable saturated porous media Driving force: pressure gradient

Source: Adapted from Ehlers, W. "Darcy, Forchheimer, Brinkman and Richards: classical hydromechanical equations and their significance in the light of the TPM," *Arch. Appl. Mech.*, 2020. https://doi.org/10.1007/s00419-020-01802-3.

Apart from Darcy's law, Brinkman equation, and Richards' equation are also used in some specific cases of porous media flow. Table 2.8 shows a brief comparison of these three equations.

In general fluid flow model, Navier-Stokes (NS) equation is commonly used. The simplified form of NS equation can be presented as follows:

$$\underbrace{\rho\frac{\partial v_z}{\partial t}}_{\text{Transient}}+\underbrace{\rho\left(v_x\frac{\partial v_z}{\partial x}+v_y\frac{\partial v_z}{\partial y}+v_z\frac{\partial v_z}{\partial z}\right)}_{\text{Inertia}}=-\underbrace{\frac{\partial p}{\partial z}}_{\text{Pressure}}+\underbrace{\mu\left(\frac{\partial^2 v_z}{\partial x^2}+\frac{\partial^2 v_z}{\partial y^2}+\frac{\partial^2 v_z}{\partial z^2}\right)}_{\text{Viscous}}+\underbrace{\rho gz}_{\text{Gravity}} \tag{2.23}$$

Fluid flow can be categorized as laminar and turbulent depending on the domination of viscosity. A more detailed discussion on fluid flow is presented in Chapter 9.

REFERENCES

1. M. U. H. Joardder, C. Kumar, and M. A. Karim, "Food structure: Its formation and relationships with other properties," *Crit. Rev. Food Sci. Nutr.*, vol. 57, no. 6, pp. 1190–1205, 2017.
2. W. K. Lewis, "The rate of drying of solid materials," *J. Ind. Eng. Chem.*, vol. 13, no. 5, pp. 427–432, 1921.
3. T. K. Sherwood, "The drying of solids—I," *Ind. Eng. Chem.*, vol. 21, no. 1, pp. 12–16, 1929.
4. N. H. Ceaglske and O. A. Hougen, "Drying granular solids," *Ind. Eng. Chem.*, vol. 29, no. 7, pp. 805–813, 1937.
5. P. S. H. Henry, "The diffusion of moisture and heat through textiles," *Discuss. Faraday Soc.*, vol. 3, pp. 243–257, 1948.
6. P. Görling, "Physical phenomena during the drying of foodstuffs. In Fundamental aspects of the dehydration of foodstuffs," *Soc. Chem. Ind.*, vol. 42, 1958.
7. W. B. Van Arsdel, *Food Dehydration*. The Avi Publishing, 1963.
8. R. B. Keey, *Drying Principles and Practice*. Pergamon Press, 1970.
9. S. Bruin, and K. C. A. M. Luyben, "Drying of food materials: a review of recent developments," in *Advances in Drying*, Vol. 1, Ed. A. S. Mujumdar. 1980.
10. B. Hallström, "Mass transport of water in foods — a consideration of the engineering aspects," *J. Food Eng.*, vol. 12, pp. 45–52, 1990.
11. K. M. Waananen, J. B. Litchfield, and M. R. Okos, "Classification of drying models for porous solids," *Dry. Technol.*, vol. 11, no. 1, pp. 1–40, 1993.
12. R. Krishna and J. A. Wesselingh, "The Maxwell-Stefan approach to mass transfer," *Chem. Eng. Sci.*, vol. 52, no. 6, pp. 861–911, 1997.
13. B. P. Van Milligen, P. D. Bons, B. A. Carreras, and R. Sanchez, "On the applicability of Fick's law to diffusion in inhomogeneous systems," *Eur. J. Phys.*, vol. 26, no. 5, p. 913, 2005.
14. L. G. Longsworth, "The mutual diffusion of light and heavy water," *J. Phys. Chem.*, vol. 64, no. 12, pp. 1914–1917, 1960.
15. H. Heinemann and J. J. Carberry, *Catalysis Reviews: Science and Engineering*. M. Dekker, 1974.
16. G. Ehrlich and K. Stolt, "Surface diffusion," *Annu. Rev. Phys. Chem.*, vol. 31, no. 1, pp. 603–637, 1980.
17. A. Kapoor, R. T. Yang, and C. Wong, "Surface diffusion," *Catal. Rev. Eng.*, vol. 31, no. 1–2, pp. 129–214, 1989.
18. G. V Kharlamov, "About diffusion mechanisms in gases and liquids from data of molecular dynamics simulation," *J. Phy. Conf. Ser*, vol. 1105, no. 1, p. 12152, 2018.
19. N. Tillmark and P. H. Alfredsson, "Experiments on transition in plane Couette flow," *J. Fluid Mech.*, vol. 235, pp. 89–102, 1992.
20. D. R. Carlson, S. E. Widnall, and M. F. Peeters, "A flow-visualization study of transition in plane Poiseuille flow," *J. Fluid Mech.*, vol. 121, pp. 487–505, 1982.
21. S. Mukhopadhyay, S. S. Roy, R. A. D'Sa, A. Mathur, R. J. Holmes, and J. A. McLaughlin, "Nanoscale surface modifications to control capillary flow characteristics in PMMA microfluidic devices," *Nanoscale Res. Lett.*, vol. 6, no. 1, p. 411, 2011.
22. D. R. Haynes, N. J. Tro, and S. M. George, "Condensation and evaporation of water on ice surfaces," *J. Phys. Chem.*, vol. 96, no. 21, pp. 8502–8509, 1992.
23. S. Basu, U. S. Shivhare, and A. S. Mujumdar, "Models for sorption isotherms for foods: A review," *Dry. Technol.*, vol. 24, no. 8, pp. 917–930, 2006.

24. M. U. H. Joardder, M. Mourshed, and M. H. Masud, *State of Bound Water: Measurement and Significance in Food Processing*. Springer, 2019.
25. Y. A. Cengel, *Heat Transfer A Practical Approach*, 5th ed. McGraw-Hill, 2016.
26. J. G. Ibarra, Y. Tao, A. J. Cardarelli, and J. Shultz, "Cooked and raw chicken meat: Emissivity in the mid-infrared region," *Appl. Eng. Agric.*, vol. 16, no. 2, p. 143, 2000.
27. M. Mahiuddin, M. I. H. Khan, N. Duc Pham, and M. A. Karim, "Development of fractional viscoelastic model for characterizing viscoelastic properties of food material during drying," *Food Biosci.*, vol. 23, pp. 45–53, 2018
28. C. Kumar, M. U. H. Joardder, T. W. Farrell, and M. A. Karim, "Investigation of intermittent microwave convective drying (IMCD) of food materials by a coupled 3D electromagnetics and multiphase model," *Dry. Technol.*, vol. 36, no. 6, pp. 736–750, 2018

3 Governing Equations and Material Properties

3.1 INTRODUCTION

Mathematical modelling in physics involves a set of steps from the proper understanding of physical phenomena to the numerical solution approach as shown in Figure 3.1. First, a clear insight into all associated physics in a phenomenon needs to be considered. The region of interest, which also known as the domain of the phenomena, then needs to be defined. It is the geometry of the system that needs to be provided in mathematical modelling. Depending on the scale of observation, the meshing operation needs to be accomplished as described in the next chapter.

Apart from these, the core of mathematical modelling is the governing equation of the associated physics and related thermophysical properties, and boundary conditions.

3.2 HEAT AND MASS TRANSFER DURING DIFFERENT TYPES OF DRYING

Although there are hundreds of drying methods applied in different materials, some common physics are involved in each of them. However, individual drying incorporates slight to moderate change in energy and mass conservation

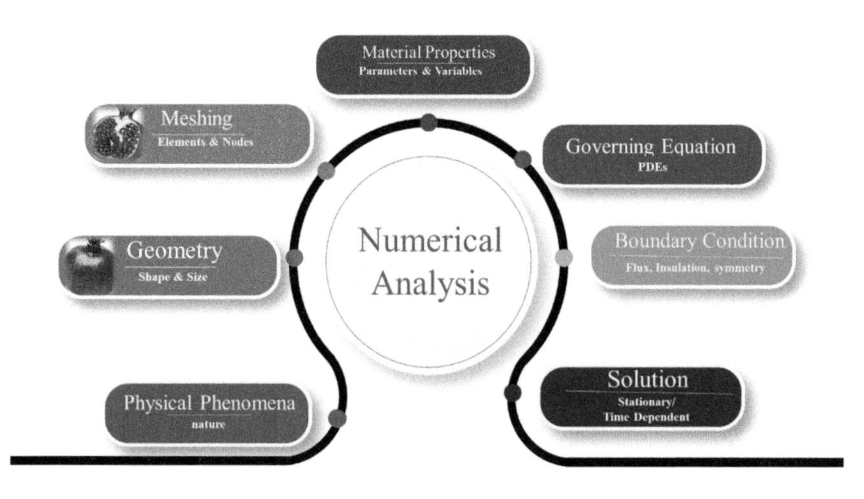

FIGURE 3.1 Essential steps of mathematical modelling.

DOI: 10.1201/9780429461040-3

equations. Hot-air drying is the most common drying method, and its involved physics of heat and mass transfer can be treated as the standard. In the following sections, we will discuss the expressing of heat and mass transfer equations for different drying systems. Heat transfer equations generally represent the thermal energy balance of the sample or drying system. The most common form of the heat transfer equations is presented in Table 3.1.

Similarly, mass transfer equations can be expressed in a different form for the convenience of application in different circumstances. The most common form of the mass balance equations is presented in Table 3.2.

In general, drying involves simultaneous heat and mass transfer. Governing equations are the same in most cases, with slight variations in expression. In the following section, we will discuss the relevant considerations when governing the different drying systems.

3.2.1 Convective Drying

Convection drying involves the convection heat and mass transfer on the exposed surface, as shown in Figure 3.2. Energy and mass balance equations can be set from Tables 3.1 and 3.2 according to set conditions.

TABLE 3.1
Some Common Forms of Heat Transport Equations [1–4]

	Form of Equation	Expression
1	The general representation for energy PDE	$\frac{\partial T}{\partial t} = \nabla \cdot (\alpha \nabla T)$
2	The 2D representation for energy PDE	$\frac{\partial T}{\partial t} = \alpha \left(\frac{\partial^2 T}{\partial x^2} + \frac{\partial^2 T}{\partial y^2} \right)$
3	The 3D representation for energy PDE	$\frac{\partial T}{\partial t} = \alpha \left(\frac{\partial^2 T}{\partial x^2} + \frac{\partial^2 T}{\partial y^2} + \frac{\partial^2 T}{\partial z^2} \right)$
4	Non-dimension scaled variables representation	$\frac{\partial \bar{T}(\zeta,\tau)}{\partial \tau} = Le \frac{\partial^2 T(\zeta,\tau)}{\partial \zeta^2}$

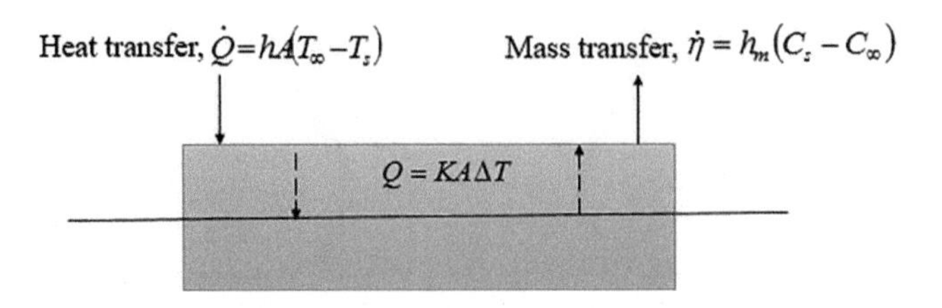

FIGURE 3.2 Heat and mass transfer mechanism during convective drying.

The main task in convection drying modelling is to find the value of the convection coefficient of heat and mass transfer. To find the approximate value of heat transfer and mass transfer coefficient, we need to consider various important factors, including air velocity, sample geometry, and sample roughness. The heat transfer coefficients are calculated from well-established correlations of the Nusselt number, as mentioned in the following equation.

$$\mathrm{Nu} = \frac{h_T L}{k} = C\,\mathrm{Re}^n\,\mathrm{Pr}^m \tag{3.1}$$

C, n, and m are the fitting parameters of the empirical relations that vary with the drying conditions. Proper values of these fitting parameters are readily available in all of the heat transfer textbooks. Similarly, the mass transfer coefficient can be determined by applying the Sherwood number (Sh), as follows:

$$\mathrm{Sh} = \frac{h_m L}{D_{va}} = C\,\mathrm{Re}^n\,Sc^m \tag{3.2}$$

The Sherwood number also needs the fitting parameters, as the Nusselt number does. The fitting constants are pretty much identical as those are used in the Nusselt number. In a controlled situation, one can use the same fitting parameter of Nusselt number. In determining the values of Nu and Sh numbers, one needs the correct values of Re, Sc, and Pr, which are calculated by the following relations:

$$\mathrm{Re} = \frac{\rho_a v L}{\mu_a}, \tag{3.3}$$

$$\mathrm{Sc} = \frac{\mu_a}{\rho_a D_{va}}, \tag{3.4}$$

$$\text{and}\,\mathrm{Pr} = \frac{C_p \mu_a}{k_a}, \tag{3.5}$$

The heat and mass transfer coefficients need to be determined for specific conditions as these vary with sample dimension, air velocity, and surface condition. Moreover, these coefficients are not temporally and spatially constant; rather, they vary with drying time and sample positions in the dryer.

3.2.2 Vacuum Drying

In vacuum drying, there is no significant difference in governing equation and boundary conditions than those of the convective drying. The single most important concern of the vacuum drying is the vacuum pressure, which is the main driving force of the moisture transfer during drying of a sample. The vapour pressure differences between the sample and the surroundings can be considered using the following boundary conditions of heat and mass transfer [7]:

TABLE 3.2
Some Common Forms of Mass Transport Equations [3–6]

	Form of Equation	Expression
1	The general representation for moisture PDE	$\frac{\partial X}{\partial t} = \nabla \cdot \left(D_{eff} \nabla X \right)$
2	The 2D representation for moisture PDE	$\frac{\partial X}{\partial t} = D_{eff} \left(\frac{\partial^2 X}{\partial x^2} + \frac{\partial^2 X}{\partial y^2} \right)$
3	The 3D representation for moisture PDE	$\frac{\partial X}{\partial t} = D_{eff} \left(\frac{\partial^2 X}{\partial x^2} + \frac{\partial^2 X}{\partial y^2} + \frac{\partial^2 X}{\partial z^2} \right)$
4	Equation expressed in cylindrical coordinates	$\frac{\partial X}{\partial t} = \frac{1}{r} \frac{\partial}{\partial r} \left(r D_{eff} \frac{\partial X}{\partial r} \right) + \frac{\partial}{\partial y} \left(D_{eff} \frac{\partial X}{\partial y} \right)$
5	The gradient in water potential	$C_m \frac{\partial \psi}{\partial t} + \nabla \cdot \left(-K_m \nabla \psi \right) = 0$

$$\mathbf{n}.(-k\nabla T) = h_T (T - T_{air}) - h_m \frac{\left(p_{v,eq} - p_{vair} \right)}{RT} h_{fg} \tag{3.6}$$

$$\left(-D_{eff} \nabla c \right) = h_m \frac{\left(p_{v,eq} - p_{vair} \right)}{RT} \tag{3.7}$$

In the above equations, $(p_{v,eq} - p_{vari})$ is the most important factor as it significantly controls simultaneous heat and mass transfer phenomena. Saturation vapour pressure is required to obtain the value of atmospheric vapour pressure at a certain relative humidity and equilibrium vapour pressure of the sample at any moisture content. Saturation vapour pressure (P_{sat}) can be obtained from the following relationships [8]:

$$P_{v,sat} = \exp \begin{bmatrix} -5800.2206/T + 1.3915 - 0.0486T + 0.4176 \times 10^{-4} T^2 \\ -0.01445 \times 10^{-7} T^3 + 6.656 \ln(T) \end{bmatrix} \tag{3.8}$$

The vapour pressure of the porous materials is assumed to always be in equilibrium with the vapour pressure given by an appropriate sorption isotherm. The relationships need to be attained from experimental observation for an individual sample. For a porous material, the correlation of equilibrium vapour pressure with moisture and temperature can be developed from sorption isotherm, and it looks like this

$$P_{v,eq} = P_{v,sat}(T) \exp\left(-aM_{db}^{-b} + ce^{-dM} M_{db}^{f} \ln[P_{sat}(T)] \right) \tag{3.9}$$

Therefore, a proper determination of equilibrium vapour pressure of a sample is vital in vacuum drying.

3.2.3 Drum Drying

Drum drying is effective for the drying of pureed materials and other high-viscous liquid materials. In the drum drying process, the drying of viscous liquid materials is applied as a thin layer onto the outer surface of drums that are continuously heated internally. The conductive–convective heat transfer approach is dominant in drum drying, and this is the main distinguishing feature of heat transfer phenomena in this type of drying. The following aspects need to be considered:

- The mass transfer takes place in one direction where the convection heat transfer occurs as well.
- The thin layer of dense liquid sample encounters two different boundary conditions. Dryer surface–thin layer of sample interface boundary condition is a specified temperature boundary condition. However, the outer layer of the sample encounters the convective boundary layer. The boundary conditions are mentioned in Section 3.3.
- Dryer surface and thin layer interface temperature stay unchanged.

3.2.4 Freeze Drying

Heat and mass transfer during freezing are associated mainly with heat conduction and diffusional mass (moisture) transfer within the solid sample, whereas convection and radiation boundary conditions prevail on the sample's exposed surface. Maintaining the appropriate pressure and temperature of the sample below those of the triple point is the main goal. In freeze drying, the solid phase of water is converted to vapour directly. At the final stage of freeze drying, evaporation takes place to facilitate water migration from the sample. Figure 3.3 shows the phase change processes during freeze drying.

In the energy conservation equation, the volumetric heat source term is often incorporated to remove the water, which cannot be removed by sublimation. The equation can be expressed as follows:

$$\rho C_p \frac{\partial T}{\partial t} = \nabla \cdot \left(k \nabla \cdot T\right) + G \tag{3.10}$$

The mass transfer equation also carries a mass generation term in the form of the sublimation rate. The Hertz–Knudsen formula is used to calculate the sublimation rate of ice during freeze drying. The following mass generation term is also reported in the literature:

$$\dot{m}_g = -\frac{\partial \theta}{\partial t} \phi \rho_I \tag{3.11}$$

where $\dot{m}$ is the rate of mass generation in kg/(m^3·s) and ρ is the ice density in kg/m^3. The general form of the heat transfer boundary condition from the surface of the solid to the external medium is the combination of convection, radiation, and evaporation.

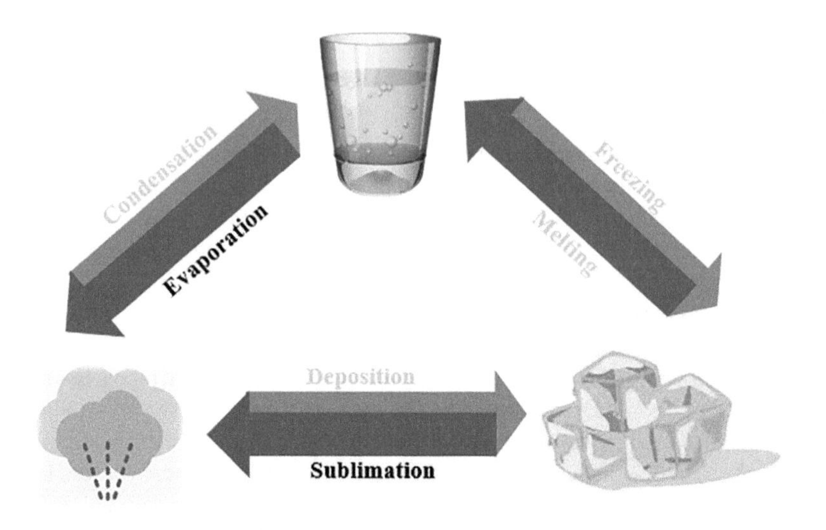

FIGURE 3.3 Phase change during freeze drying.

$$-K\frac{dT}{dx} = \text{convection} + \text{radiation} + \text{evaporation} \tag{3.12}$$

The main concerns in developing freeze-drying modelling are the rate equation in the form of source/sink term of the governing equation and as a boundary condition.

3.2.5 Spray Drying Process

The spray drying process is the multiphase drying problem comprising a gas phase (drying air and water vapour), a liquid phase (droplets of drying materials), and a solid phase (dried particles). Both continuous phases and discrete phases need to be considered in spray drying modelling. The gas-phase is considered to be a continuous phase, whereas the liquid and solid phases are considered as discrete phases. The conservation equations and balance equation of transport phenomena, which are required for modelling spray drying, are presented in Table 3.3.

The above equations are often applied in spray drying modelling where the complex phenomena such as particle–particle and droplet–droplet collisions are often neglected.

3.2.6 Microwave Drying

Microwave (MW) causes heat production in the presence of water molecules in the sample [15]. In MW drying modelling, heat generation due to MW is considered as an energy source term. The heat generation due to MW can be inncluded using Lambert's equation or Maxwell's equation. Both equations have advantages

TABLE 3.3
Governing Equations for Spray Drying Modelling [9–14]

	Conservation Equation	Expression
Continuous phase	Mass conservation	$\frac{\partial \rho}{\partial t}+\frac{\partial}{\partial x_j}(\rho u_j)=Sc$
	Momentum conservation	$\frac{\partial}{\partial t}(\rho u_i)+\frac{\partial}{\partial x_j}(\rho u_j u_j)=-\frac{\partial p}{\partial x_j}+\frac{\partial}{\partial x_j}\left[\mu_e\left(\frac{\partial u_i}{\partial x_j}+\frac{\partial u_j}{\partial x_j}\right)\right]+\Delta\rho g_i+U_{pi}S_c+\sum F_{gp}$
	Energy conservation	$\frac{\partial}{\partial t}(\rho h)+\frac{\partial}{\partial x_j}(\rho u_j h)=\frac{\partial}{\partial x_j}\left(\frac{\mu_e}{\sigma_h}\frac{\partial h}{\partial x_j}\right)-q_r+S_h$
	Turbulence kinetic energy	$\frac{\partial}{\partial t}(\rho k)+\frac{\partial}{\partial x_j}(\rho u_j k)=\frac{\partial}{\partial x_j}\left(\frac{\mu_e}{\sigma_k}\frac{\partial k}{\partial x_j}\right)+G_k+G_b-\rho\varepsilon$
Discrete phase	Droplet momentum conservation	$\frac{d\vec{U}_p}{dt}=\vec{g}+\frac{\sum \vec{F}_p}{m_p}$ where the total force, $\sum \vec{F}_p=\vec{F}_D+\vec{F}_A+\vec{F}_B+\vec{F}_C$ comprises body force, contact force, drag force and buoyancy force

and limitations depending on the perspective [16]. Maxwell's equation is more accurate in modelling microwave distribution and heat generation. However, it is far more computationally expensive than Lambert's approach.

3.2.6.1 Maxwell's Equation for Electromagnetics

Maxwell's equations provide the electromagnetic field distribution at any point in the computational domain. In the frequency domain, Maxwell's equation can be written as [17]:

$$\nabla\times\left(\frac{1}{\mu}\nabla\times\vec{E}\right)-\frac{(2\pi f)^2}{c}(\varepsilon'-i\varepsilon'')\vec{E}=0 \tag{3.13}$$

where $\vec{E}$ is the electric field strength (V/m); f is the microwave frequency (Hz); c is the speed of light (m/s); and ε', ε'', and μ are the dielectric constant, dielectric loss factor, and electromagnetic permeability of the material, respectively.

The heat generation due to microwave, Q_m (W/m^3), is given by Chen et al. [17]. Substituting the above equation, the microwave heat generation due to can be written as:

$$Q_m=\pi f\varepsilon_0\varepsilon''|E|^2 \tag{3.14}$$

3.2.6.2 Lambert's Law

Lambert's law can be used to calculate microwave energy absorption inside the food samples. Lambert's law considers exponential attenuation of microwave absorption within the product, by the following relationship:

$$P(x) = P_0 \exp^{(-2\alpha x)} \quad (3.15)$$

where P_0 is the incident power at the surface and α is the attenuation constant that can be expressed with the following equations:

$$\alpha = \frac{2\pi}{\lambda}\sqrt{\varepsilon' \left[\frac{\sqrt{\left(1+\left(\frac{\varepsilon''}{\varepsilon'}\right)^2\right)}-1}{2}\right]} \quad (3.16)$$

The overall absorption coefficient depends on the infrared emitter emissivity, the emissivity of the sample, and the shape factor between the infrared emitter and the sample.

3.2.7 Infrared Drying

Infrared (IR) carries thermal radiation, and irradiation of this on a surface causes heating of the surface. A surface with high emissivity is suitable for infrared drying. Heat transfer from the heating source to the product surface spontaneously occurs without interfering with the surrounding air. The wavelength of infrared heating and radiation properties of the surface, including absorptivity, transitivity, and emissivity, are the main factors that affect the heat transfer rate during IR drying. Along with the volumetric infrared heat source (Q), the energy conservation equation for IR drying can be written as follows [18]:

$$\rho c_p \frac{\partial T}{\partial t} + \rho c_p u \cdot \nabla T = (k\nabla T) + \dot{Q} \quad (3.17)$$

Lambert's law quantifies the infrared heat generated in terms of heat absorbed by the product. The amount of heat penetrating a product decays exponentially in relation to the coefficient of absorption and the depth of penetration as mentioned in the preceding section. Moreover, the following expression can be used in the simplified modelling of IR drying:

$$\dot{Q} = \sigma A_s \left(\varepsilon T_s^4 - \alpha T_{sur}^4\right) \quad (3.18)$$

3.2.8 Combined or Assisted Drying

In the case of a combined drying system, two or more drying techniques are combined to obtain efficient drying. All of the possible heat and mass transfer of such combined systems must be taken into consideration, either in governing equation or in the boundary equations. Combined drying systems can be either continuous or intermittent.

3.2.8.1 Continuous

There are some drying systems which need assistance from other drying systems continuously. Freeze drying is often combined with vacuum and IR drying systems [19]. Convection drying is also incorporated in many drying systems including MW drying. The general energy, mass, and momentum conservation must be ensured in all of the combined or assisted drying. Special consideration needs to be taken with a combined drying system, which can be seen as follows:

- Governing equations must account for all of the energy provided to the sample
- Energy generation in case of IR and MW needs to be considered as the source term of the energy conservation equation.
- All of the internal mass transfer equations including diffusion and external mass transfer mechanisms including convection and sublimation need to be incorporated accordingly.

In many cases, additional energy sources are supplied intermittently to achieve optimum energy consumption.

3.2.8.2 Intermittent

Intermittent drying means addition of energy source or controlling drying parameters in an intermittent pattern. Intermittency can be achieved in different pathways. Figure 3.4 shows the possible way of intermittent drying [20].

The intermittent term of the energy needs to be added in the governing equation. For instance, in intermittent convective microwave drying, microwave heating, Q_{mic} (W/m^3) is added in an intermittent pattern and expressed as follows:

$$\rho c_p \frac{\partial T}{\partial t} = \nabla.(k\nabla T) + Q_{mic} f(t) \quad (3.19)$$

Heat generation Q_{mic} (W/m^3) is incorporated in an interval by using a function $f(t)$. In other words, MW is ON for a short while and set OFF for the rest of the time in a cycle. The ratio of ON time and the total time of the cycle is defined as pulse ratio (PR).

3.3 BOUNDARY CONDITIONS

Boundary conditions are essential to consider the effect of the surrounding on the sample. Separate boundary conditions are required on each boundary of the sample. A 1D problem needs two boundary conditions, whereas a 2D problems

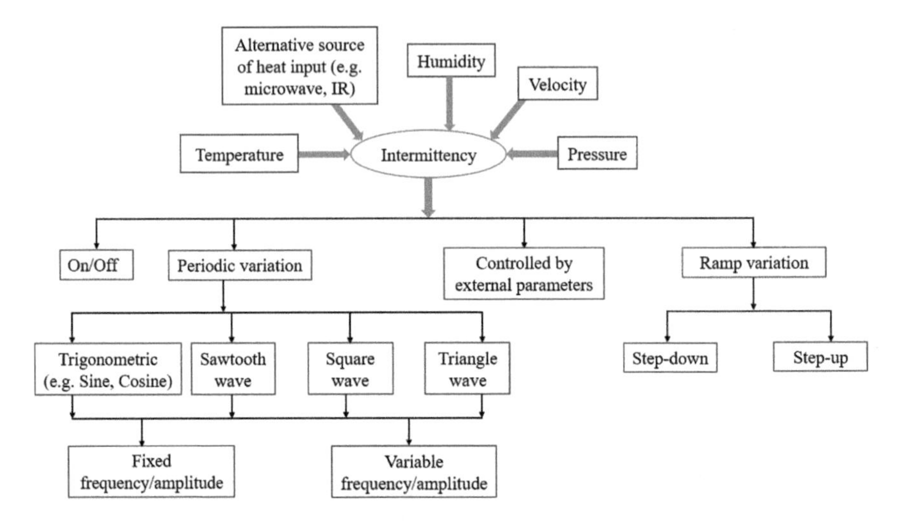

FIGURE 3.4 Possible options for achieving intermittency in drying. (Adapted from [20].)

needs 4 boundary conditions. The most frequent boundary conditions for heat and mass transfer are discussed briefly in the following section.

3.3.1 Heat Transfer Boundary Conditions

There are many ways to define the boundary of a system. Specified surface temperature, specified heat flux, convection heat transfer, and radiation heat transfer are the most common boundary conditions for heat transfer problems as shown in Figure 3.5.

The generalized boundary condition equation for heat transfer must follow the relations as shown in Table 3.4.

- **Specified surface temperature** is the simplest boundary condition. The boundary temperature is kept constant throughout the drying system. It is a Dirichlet condition as specified temperature boundary condition generally remains constant. The constant temperature can be maintained in a conduction interface between a surface of the dryer and a boundary of the sample.
- **Specified heat flux boundary** is a function of space and/or time and it specified as a constant. There are two special heat flux boundary conditions, namely insulation boundary condition and symmetry boundary condition. If there is no heat flux across the surface $\frac{\partial T}{\partial x} = 0$, the boundary can be specified as insulation. Zero heat flux $\left(\frac{\partial T}{\partial x} = 0\right)$ is

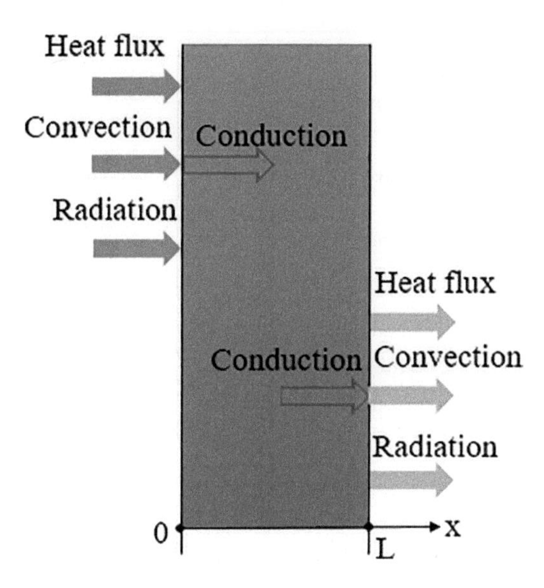

FIGURE 3.5 Possible heat transfer boundary conditions during drying of a sample.

TABLE 3.4
Heat Transfer Boundary Conditions

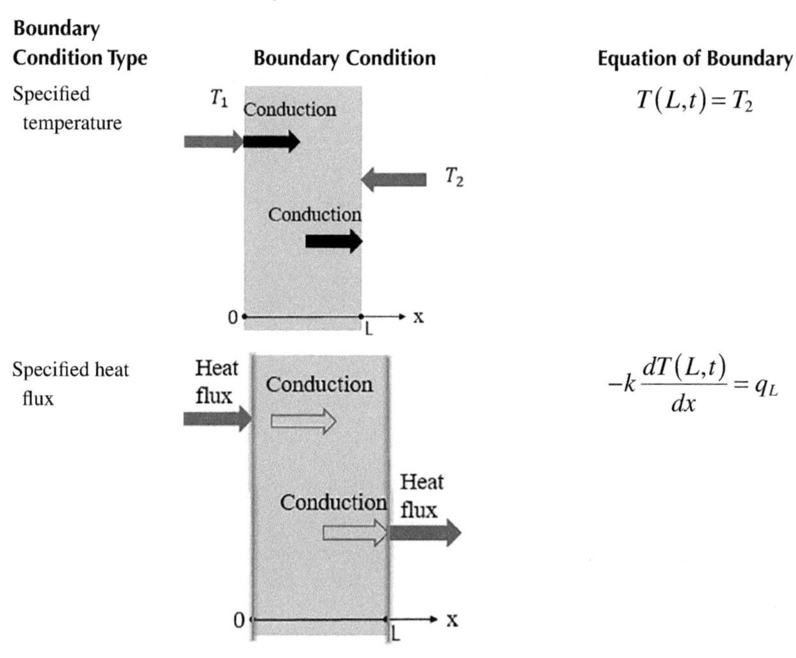

Boundary Condition Type	Boundary Condition	Equation of Boundary
Specified temperature		$T(L,t) = T_2$
Specified heat flux		$-k\dfrac{dT(L,t)}{dx} = q_L$

(Continued)

TABLE 3.4 (*Continued*)
Heat Transfer Boundary Conditions

Boundary Condition Type	Boundary Condition	Equation of Boundary
Insulated boundary		$\frac{dT(L,t)}{dx} = 0$
Symmetry at a boundary		$\frac{dT\left(\frac{L}{2},t\right)}{dx} = 0$
Convection at boundary		$-K\frac{dT(L,t)}{dx} = h_2\,[T_s - T_\infty]$
Radiation at boundary		$-k\frac{dT(L,t)}{dx} = \varepsilon_2\sigma[T^4 - T^4_{surr,2}]$

Source: Adapted from multiple sources including [21].

also maintained in a boundary, which is the centre line due to symmetry of the sample.

- **The convection boundary condition** is used when there is a fluid motion across a boundary of the sample. Most of the boundary of the sample in the convective dryer maintains this boundary condition.
- **Radiation boundary conditions** are used when irradiation on a boundary is caused by a remote heating source. Apart from the significant heating source, the radiation boundary condition is often neglected to simplify the boundary conditions. On the other hand, the radiation boundary condition may merge with convective boundary condition and the boundary conditions look like the following equation:

$$-k \cdot \left.\frac{\partial T}{\partial x}\right|_{\sigma} = h \cdot \left(T|_{\sigma} - T_{\infty}\right) + \varepsilon \cdot \sigma\left(T|_{\sigma}^{4} - T_{\infty}^{4}\right)c \tag{3.20}$$

3.3.2 Mass Transfer Boundary Conditions

Similar to heat transfer, the mass transfer problem needs appropriate boundary conditions. Although the governing equation of mass transfer is quite analogous with heat transfer, mass transfer boundary conditions have some specific features. For example, the concentration of a species on the two sides of a multiphase interface (liquid–gas or solid–gas or solid–liquid) are different, whereas temperatures in such conditions remain the same. The common types of mass transfer boundary conditions, which are analogous to those of heat transfer, are as follows:

- **The specified concentration of a species ($C = C_1$)** is similar to specified temperature boundary conditions. This boundary condition is possible in a controlled environment where the concentration of a certain species remains constant.
- **Specified mass flux** can be maintained using a boundary source term of a targeted species. There are two special mass flux boundary conditions, namely, zero-flux boundary condition and symmetry boundary condition. If there is no mass flux across the surface $\frac{\partial C}{\partial x} = 0$, the boundary can be specified as a zero-flux boundary. No mass flux $\left(\frac{\partial C}{\partial x} = 0\right)$ is also maintained in a boundary that is the centre line due to symmetry of the sample.
- **The convection boundary condition** is used when there is a fluid motion across a boundary of the sample. Similar to the heat transfer problem, convection boundary conditions are common boundary conditions during drying of a sample where airflow is involved.

After discussing the governing equation and boundary condition, we will concentrate on the properties of the common materials related to heat and mass transfer.

It is not the purpose of this section to present the properties in details. Instead, this chapter is devoted to a brief review of the main properties required in developing heat and mass transfer modelling.

3.4 THERMO-PHYSICAL AND TRANSPORT PROPERTIES

Thermo-physical and transport properties are significantly important in achieving proper distribution of heat and mass in a sample during drying. Thermal diffusivity and effective moisture diffusivity are the most important transport properties that directly or indirectly depend on the thermo-physical properties of the materials [22,23]. Therefore, we will present the thermo-physical properties along with the mentioned transport properties. Before entering into the discussion of the individual properties, it is important to mention some of the common concerns of the properties.

- In general, the properties are not spatially or temporally constant; rather, these vary with moisture content and with drying time. For simplification, the values of the properties are often considered as constant.
- It is very difficult to find all of the properties for most of the moisture-rich materials in the literature. On the other hand, an experimental determination is not feasible in many situations. Unavailable or indeterminable properties of materials can be taken from those of similar materials for modelling purposes.
- The effective value of the properties can be obtained from the empirical relationship developed by considering the constituents found in the sample. For example, water, carbohydrate, and protein are the common constituents of food materials, and the properties of these constituents can be used in determining the properties of the selected materials.

3.4.1 Density

Density, shrinkage, and porosity are the mass- and volume-related properties that constitute one of the main groups of mechanical properties [24]. Density can be found from the information of the volume and mass of the sample. However, determination of density for an irregularly shaped sample is not a simple task due to complexity in the calculation of the volume of the sample.

3.4.2 Porosity

Porous materials are a combination of solid, liquid, and gas. Porosity is the void volume of air in the total volume. It can be obtained from the following relationship:

$$\varepsilon = \frac{V_a}{V_s + V_w + V_a} = 1 - \frac{V_s + V_w}{V} \tag{3.21}$$

Porosity can be estimated from the apparent density and the true density of the material, and thus it can be alternatively expressed according to the following equation:

$$\varepsilon = 1 - \frac{\rho_b}{\rho_p} \tag{3.22}$$

The drying of any porous products affects the structural orientation because of the simultaneous heat and mass transfer. Pore formation and evolution is one of the important structural changes that occur during drying. However, the mechanisms of pore formation in most materials throughout drying are very complicated because of structural complexity. In general, void development because of water migration and structural mobility are more pronounced phenomena that cause pore formation and evolution during drying.

There are two approaches to predicting porosity of a sample during drying. Empirical models are developed using experimental data to fit parameters for a particular process. Methods, materials, and the processing environment limit these models, and the fitting parameters usually have no physical significance. Despite the existence of these limitations, the empirical models are beneficial to a certain extent.

On the other hand, the theoretical model conveys the physics associated with pore development. Most of the theoretical models are based on the conversion of mass and volume principle. The first theoretical porosity model was developed by Kilpatric et al. [25]. It was a simple model that considered the volumetric shrinkage of fruits and vegetables during drying. This model only considers bulk and solid density, as shown in Equation (3.21) [26].

$$\varepsilon = 1 - \frac{\rho_b}{\rho_s} \tag{3.23}$$

Although this model has physical significance, it needs instantaneous experimental values of bulk density of the sample. Lozano et al. [24] developed the following generalized equation to predict the porosity of foodstuffs. It assumes that a linear relationship between porosity and moisture content exists during drying and is expressed by:

$$\varepsilon = \frac{(X+1)}{(X_0+1)} \frac{\rho_{b0}}{\rho_b} \tag{3.24}$$

There are many other theoretical models developed in recent years. For more details of the porosity evolution during drying, refer to the work of Joardder et al. [23,27].

3.4.3 Specific Heat Capacity

Specific heat capacity refers to the amount of heat energy a material can store. It plays a vital role in distributing thermal energy within a material. A material with low specific heat capacity encounters fast temperature rise. As most of the materials subject to drying have a substantial amount of water, the value of specific heat capacity of fresh/raw materials is close to that of water. There are some models used to predict the specific heat capacity of different materials with different moisture content. The common equation for calculating the specific heat capacity of different food materials can be predicted by the following equation [28]:

$$c_p = 0.4 + (0.6 \times \text{water content}) \tag{3.25}$$

Specific heat also varies with temperature; the change is remarkable at the freezing point, as shown in Table 3.5.

3.4.4 Thermal Conductivity

Thermal conductivity is defined as the ability of a material to conduct heat. The structure of the foodstuff, its composition, heterogeneity of food, and processing conditions influence the thermal conductivity of foods. Dry porous solids are very poor heat conductors because the pores are occupied by air. For porous materials, the measured thermal conductivity is an apparent one, called the effective thermal conductivity. It is an overall thermal transport property assuming that heat is transferred by conduction through the solid and the porous phase of the material.

There are many models of thermal conductivity for heterogeneous materials available, as shown in Table 3.6. These are the models generally used to calculate the thermal conductivity of a heterogeneous material composed of a mixture of several constituents with known thermal conductivities and volume fractions.

The structural factors, including pore characteristics and distribution of different phases such as air, water, ice, and solids, influence the overall thermal conductivity of food materials. Furthermore, processing factors including temperature, pressure, and the mode of heat or energy transfer significantly influence thermal conductivity of a material. In any case, thermal diffusivity is used instead

TABLE 3.5
Specific Heat Capacity of Common Drying Fruits at Different Conditions [29]

Sl. No.	Fruits	Constant Values of Specific Heat	Specific Heat above Freezing Point (J/kg-K)	Specific Heat below Freezing Point (J/kg-K)
01.	Apple	3,724–4,017 (J/kg-K)	3,650	1.90
02.	Grape	3,703 (J/kg-K)	3,590	1,870
03.	Lemon	3,732 (J/kg-K)	3,820	1,960
04.	Orange	3,661 (J/kg-K)	3,750	1,940
05.	Plum	3,514 (J/kg-K)	3,720	1,920

TABLE 3.6
Effective Thermal Conductivity Models for Heterogeneous Materials [30–32]

Model	Thermal Conductivity Equation	Nomenclature
Parallel	$k_e = \sum_i k_i v_i$	k_e = Effective conductivity k_c = Conductivity of condensed material
Series	$k_e = \sum_i \frac{1}{\frac{r_i}{k_i}}$	k_a = Conductivity of air V_c = Volume of condense material V_c = Volume of air
M-E1 (Maxwell-Eucken, air dispersed)	$k_e = k_c \frac{2k_c + k_a - 2(k_c - k_a)\varepsilon}{2k_c + k_a + (k_c - k_a)\varepsilon}$	ε = Porosity
M-E2 (Maxwell-Eucken, condensed phase dispersed)	$k_e = k_a \frac{2k_a + k_c - 2(k_a - k_c)(1-\varepsilon)}{2k_a + k_c + (k_a - k_c)(1-\varepsilon)}$	
EMT (effective medium theory)	$v_c \frac{k_c - k_e}{k_c + 2k_e} + v_a \frac{k_a - k_e}{k_a + 2k_e} = 0$	

of thermal conductivity. Thermal diffusivity indicates how fast heat propagates through a sample during heating. Thermal diffusivity can be defined as the following expression:

$$\alpha = \frac{k}{\rho C_p} \tag{3.26}$$

From the above equation, it is depicted that thermal diffusivity can be determined from the known values of density, conductivity, and specific heat capacity of the material.

3.4.5 Emissivity

Emissivity is a measure of how much radiation is emitted from a surface. The value of emissivity of a surface is denoted by ε, which shows the values $0 \le \varepsilon \le 1$. Emissivity is a very important property in the drying where radiation is associated directly, such as IR drying. In general, thermal radiation is a volumetric phenomenon for transparent materials. However, for opaque solids, including foods, wood, and brick, radiation can be considered as a surface phenomenon. The low energy content of thermal radiation of the inner regions of materials is not sufficient enough to reach the surface. Therefore, surface properties such as emissivity, reflectivity and absorptivity of sample need to be considered prior to choosing thermal radiation associated drying. The emissivities of some materials of interest in drying are given in Table 3.7.

TABLE 3.7
Emissivity of Some Selected Materials [21,33]

Material	Emissivity	Remarks
Water	0.95–0.96	The emissivity of the materials are measured at 300 K
Ice	0.97	
Beef	0.73–0.74	
Vegetation	0.92–0.96	
Paper	0.92–0.97	
Wood	0.82–0.92	
Soil	0.93–0.96	

3.4.6 Dielectric Properties

Dielectric property is essential in modelling heat and mass transfer of MW drying. The dielectric property of a material determines the degree of interaction between the applied microwave field and the sample. These properties determine the heating rate of the products by microwave power. These properties are significantly influenced by the moisture content, composition of the food, and to some extent, the structure of foodstuff. Porous material containing air shows lower dielectric properties as air has a relative dielectric constant of 1 and a loss factor of 0. Therefore, the more porous the material becomes during drying, the more entrapped air is present in it, causing low dielectric properties of the material. In addition, the temperature and frequency of MW affect the dielectric properties. Moreover, moisture content and temperature of the sample are the most important parameters.

There are many empirical relationships of dielectric properties of different materials in literature. For example, the empirical relationship of dielectric constant and dielectric loss factors for potato are reported as follows [34,35]:

$$\varepsilon' = 64.5876 + 0.0056M - 0.2223T - 0.0046MT \tag{3.27}$$

$$\varepsilon'' = -8.2227 + 0.2360M + 0.2041T \tag{3.28}$$

3.4.7 Effective Moisture Diffusivity

The complex nature of porous materials causes more than one internal mass transfer mechanism during the time of drying. Different modes of mass transfer may exist at the same time in a system [36]. However, one of them dominates at a certain time in the system. For example drying of porous food materials such as apple predominantly follows the falling rate period. Furthermore, diffusivity is commonly used to describe drying kinetics of porous materials, including plant tissue in their falling rate stage, and the driving force of diffusion is concentration gradient [37]. This falling rate period of drying can be modelled using Fick's law and can be expressed as:

$$\frac{\partial MR}{\partial t} = \nabla\left[D_{eff}\left(\nabla MR\right)\right] \tag{3.29}$$

where moisture ratio, MR, is calculated using the following equation:

$$MR = \frac{M - M_e}{M_0 - M_e} \tag{3.30}$$

For one-directional drying in an infinite slab, Crank [38] gave an analytical solution, as given below:

$$\text{MR} = \frac{8}{\pi^2}\sum_{n=0}^{\infty}\frac{1}{(2n+1)^2}\exp\left(-\frac{(2n+1)^2\pi^2 D_{eff}t}{4L^2}\right) \tag{3.31}$$

where n is a positive integer, t is drying time (seconds), and L is sample thickness (m). Considering uniform initial moisture distribution and negligible shrinkage, Equation (3.29) is suitable for determining effective moisture diffusivity. A simplified approach is shown in Equation (3.30) and can be obtained taking only the first term of series solutions in Equation (3.30). This equation could be further simplified into Equation (3.31) by taking the first term of a series solution as follows:

$$\ln\left(\frac{M - M_e}{M_o - M_e}\right) = \ln\left(\frac{8}{\pi^2}\right) - \left(\frac{\pi^2 D_{eff}}{4L^2}t\right) \tag{3.32}$$

$$\ln\left(\frac{M - M_e}{M_o - M_e}\right) = \ln\left(\frac{8}{\pi^2}\right) - Kt \tag{3.33}$$

Effective moisture diffusion could be determined from the slope (k) obtained from the plot of $\ln\left(\frac{M - M_e}{M_o - M_e}\right)$ versus time. The slope of the straight line can be expressed as follows:

$$\text{Slope} = \frac{\pi^2 D}{4L^2} = K \tag{3.34}$$

Putting the value of K for different shapes as given in Table 3.8, the value of effective moisture diffusivity can easily be calculated.

As diffusivity depends on the temperature of the drying system, the temperature-dependent effective diffusivity can be calculated from the Arrhenius-type relationship, as shown below:

TABLE 3.8
Effective Moisture Diffusivity Determination [39–42]

Shapes	Effective Moisture Diffusivity Equations	Slope
	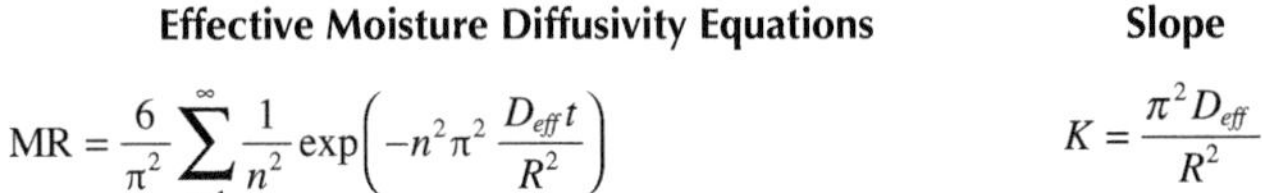 $\mathrm{MR} = \frac{6}{\pi^2}\sum_{n=1}^{\infty}\frac{1}{n^2}\exp\left(-n^2\pi^2\frac{D_{eff}t}{R^2}\right)$	$K = \frac{\pi^2 D_{eff}}{R^2}$
	$\mathrm{MR} = \frac{M - M_e}{M_0 - M_e} = \sum_{n=1}^{\infty}\frac{4}{r^2(\alpha_n)^2}\exp\left(-(\alpha_n)^2\frac{Dt}{r^2}\right)$	$K = \frac{D_{eff}\alpha_1^2}{r^2}$
	$\mathrm{MR} = \frac{M - M_e}{M_0 - M_e} = \frac{8}{\pi^2}\sum_{n=1}^{\infty}\frac{1}{(2n-1)^2}\exp\left(-\frac{(2n-1)^2\pi^2 D_{eff}t}{4L^2}\right)$	$K = \frac{\pi^2 D_{eff}}{4L^2}$
	$\mathrm{MR} = \frac{512}{\pi^6}\left[\sum_{i=1}^{\infty}\frac{1}{(2i-1)^2}\exp\left(-(2i-1)kt\right)^2\right]^3$	$K = \frac{4\pi^2 D_{eff}}{L^2}$

$$D_{eff}(T) = D_o \exp\left(-\frac{E}{R_g T}\right) \quad (3.35)$$

where, D_o is the Arrhenius factor (m^2/s), E is Activation energy (kJ/mol), and R_g is molar gas constant (kJ/mol/K). In most cases, diffusivity is affected by both sample properties and process parameters. Process parameters can easily be controlled compared to sample properties. The effective diffusivity varies throughout the process due to moisture and temperature gradient. For average diffusivity, the following expression can be used [43,44]:

$$D_{eff}(M,T) = \frac{\left(D(M) + D(T)\right)}{2} \quad (3.36)$$

The above-mentioned properties of the material are directly related to the governing physics of heat and mass transfer. The accurate values of these parameters are vital in differentiating temperature and moisture distribution within the samples.

These are the parameters that demonstrate the materials in the modelling. Many of the parameters of different materials are readily available, whereas the distinct nature of individual materials demands the experimental determination of the values of the parameters.

REFERENCES

1. R. Golestani, A. Raisi, and A. Aroujalian, "Mathematical modeling on air drying of apples considering shrinkage and variable diffusion coefficient," *Dry. Technol.*, vol. 31, no. 1, pp. 40–51, 2013.
2. J. A. Esfahani, H. Majdi, and E. Barati, "Analytical two-dimensional analysis of the transport phenomena occurring during convective drying: apple slices," *J. Food Eng.*, vol. 123, pp. 87–93, 2014.
3. V. P. C. Mohan and P. Talukdar, "Three dimensional numerical modeling of simultaneous heat and moisture transfer in a moist object subjected to convective drying," *Int. J. Heat Mass Transf.*, vol. 53, no. 21–22, pp. 4638–4650, 2010.
4. N. Shahari, N. Jamil, and K. A. Rasmani, "Comparative study of shrinkage and non-shrinkage model of food drying," *J. Phys. Conf. Ser.*, vol. 738, p. 12087, 2016.
5. R. Baini and T. A. G. Langrish, "Choosing an appropriate drying model for intermittent and continuous drying of bananas," *J. Food Eng.*, vol. 79, no. 1, pp. 330–343, 2007.
6. W. Aregawi et al., "Understanding forced convective drying of apple tissue: Combining neutron radiography and numerical modelling," *Innov. Food Sci. Emerg. Technol.*, vol. 24, pp. 97–105, 2014.
7. M. Murru, G. Giorgio, S. Montomoli, F. Ricard, and F. Stepanek, "Model-based scale-up of vacuum contact drying of pharmaceutical compounds," *Chem. Eng. Sci.*, vol. 66, no. 21, pp. 5045–5054, 2011.
8. H. Vega-Mercado, M. M. Góngora-Nieto, and G. V Barbosa-Cánovas, "Advances in dehydration of foods," *J. Food Eng.*, vol. 49, no. 4, pp. 271–289, 2001.
9. Z. Lixing and A. Arbel, "Theory and numerical modeling of turbu-lent gas-particle flows and combustion," *Combustion*, vol. 7721, p. 8, 1995.
10. L.-S. Fan and C. Zhu, *Principles of Gas-Solid Flows*. Cambridge University Press, 2005.
11. C. T. Crowe, J. D. Schwarzkopf, M. Sommerfeld, and Y. Tsuji, *Multiphase Flows with Droplets and Particles*. CRC Press, 2011.
12. M. Mezhericher, A. Levy, and I. Borde, "Droplet–droplet interactions in spray drying by using 2D computational fluid dynamics," *Dry. Technol.*, vol. 26, no. 3, pp. 265–282, 2008.
13. M. Mezhericher, *Drying of Slurries in Spray Dryers*. Ben Gurion University, 2008.
14. M. Mezhericher, A. Levy, and I. Borde, "Modeling of droplet drying in spray chambers using 2D and 3D computational fluid dynamics," *Dry. Technol.*, vol. 27, no. 3, pp. 359–370, 2009.
15. M. U. H. Joardder, M. H. Masud, S. Nasif, J. A. Plabon, and S. H. Chaklader, "Development and performance test of an innovative solar derived intermittent microwave convective food dryer," *AIP Conf. Proc.*, vol. 2121, no. 1, p. 40010, 2019.
16. C. Kumar, M. U. H. Joardder, T. W. Farrell, and M. A. Karim, "Investigation of intermittent microwave convective drying (IMCD) of food materials by a coupled 3D electromagnetics and multiphase model," *Dry. Technol.*, vol. 36, no. 6, pp. 736–750, 2018.

17. J. Chen, K. Pitchai, S. Birla, M. Negahban, D. Jones, and J. Subbiah, "Heat and mass transport during microwave heating of mashed potato in domestic oven—Model development, validation, and sensitivity analysis," *J. Food Sci.*, vol. 79, no. 10, pp. E1991–E2004, 2014.
18. D. I. Onwude, N. Hashim, K. Abdan, R. Janius, G. Chen, and C. Kumar, "Modelling of coupled heat and mass transfer for combined infrared and hot-air drying of sweet potato," *J. Food Eng.*, vol. 228, pp. 12–24, 2018.
19. Y.-P. Lin, T.-Y. Lee, J.-H. Tsen, and V. A.-E. King, "Dehydration of yam slices using FIR-assisted freeze drying," *J. Food Eng.*, vol. 79, no. 4, pp. 1295–1301, 2007.
20. C. Kumar, M. A. Karim, and M. U. H. Joardder, "Intermittent drying of food products: A critical review," *J. Food Eng.*, vol. 121, pp. 48–57, 2014.
21. Y. A. Cengel, S. Klein, and W. Beckman, *Heat Transfer: A Practical Approach*, vol. 141. WBC McGraw-Hill, 1998.
22. M. U. H. Joardder, M. Mourshed, and M. H. Masud, *State of Bound Water: Measurement and Significance in Food Processing.* Springer, 2019.
23. M. U. H. Joardder, A. Karim, C. Kumar, and R. J. Brown, *Porosity: Establishing the Relationship between Drying Parameters and Dried Food Quality.* Springer, 2015.
24. M. Mahiuddin, M. I. H. Khan, N. Duc Pham, and M. A. Karim, "Development of fractional viscoelastic model for characterizing viscoelastic properties of food material during drying," *Food Biosci.*, vol. 23, pp. 45–53, 2018
25. P. W. Kilpatrick, E. Lowe, and W. B. Van Arsdel, "Tunnel dehydrators for fruit and vegetables," *Adv. Food Res.*, vol. 6, no. 360, pp. 60123–60126, 1955.
26. C. A. Miles, G. van Beek, and C. H. Veerkamp, "Calculation of thermophysical properties of foods," In *Physical Properties of Foods*, Eds. R. Jowitt, F. Escher, B. Hallstr¨om, H.F.Th. Meffert, W.E.L. Spiess and G. Vos. Applied Science Publishers, pp. 269–312, 1983.
27. M. U. H. Joardder, C. Kumar, and M. A. Karim, "Prediction of porosity of food materials during drying: Current challenges and directions," *Crit. Rev. Food Sci. Nutr.*, vol. 58, no. 17, pp. 2896–2907, 2018.
28. Y. A. Cengel, *Heat and Mass Transfer.* Tata McGraw-Hill Education, 2007.
29. M. S. Rahman, *Food Properties Handbook.* CRC Press, 2009.
30. M. Zielinska, E. Ropelewska, and M. Markowski, "Thermophysical properties of raw, hot-air and microwave-vacuum dried cranberry fruits (Vaccinium macrocarpon)," *LWT-Food Sci. Technol.*, vol. 85, pp. 204–211, 2017.
31. S. Torquato and H. W. Haslach Jr, "Random heterogeneous materials: microstructure and macroscopic properties," *Appl. Mech. Rev.*, vol. 55, no. 4, pp. B62–B63, 2002.
32. J. K. Carson, S. J. Lovatt, D. J. Tanner, and A. C. Cleland, "Thermal conductivity bounds for isotropic, porous materials," *Int. J. Heat Mass Transf.*, vol. 48, no. 11, pp. 2150–2158, 2005.
33. G. F. Davies, C. M. D. Man, S. D. Andrews, A. Paurine, M. G. Hutchins, and G. G. Maidment, "Potential life cycle carbon savings with low emissivity packaging for refrigerated food on display," *J. Food Eng.*, vol. 109, no. 2, pp. 202–208, 2012.
34. C. Kumar, M. U. H. Joardder, T. W. Farrell, and M. A. Karim, "Multiphase porous media model for intermittent microwave convective drying (IMCD) of food," *Int. J. Therm. Sci.*, vol. 104, pp. 304–314, 2016.
35. C. Kumar, M. U. H. Joardder, T. W. Farrell, G. J. Millar, and M. A. Karim, "Mathematical model for intermittent microwave convective drying of food materials," *Dry. Technol.*, vol. 34, no. 8, pp. 962–973, 2016.

36. M. U. H. Joardder, R. Alsbua, W. Akram, and M. A. Karim, “Effect of sample rugged surface on energy consumption and quality of plant-based food materials in convective drying,” *Dry. Technol.*, pp. 1–10, 2020. DOI:10.1080/07373937.2020.1745824
37. M. U. H. Joardder, A. Karim, C. Kumar, and R. J. Brown, “Determination of effective moisture diffusivity of banana using thermogravimetric analysis,” *Procedia Eng.*, vol. 90, pp. 538–543, 2014.
38. A. M. Castro, E. Y. Mayorga, and F. L. Moreno, “Mathematical modelling of convective drying of fruits: A review,” *J. Food Eng.*, vol. 223, pp. 152–167, 2018.
39. İ. Doymaz and O. İsmail, “Drying characteristics of sweet cherry,” *Food Bioprod. Process.*, vol. 89, no. 1, pp. 31–38, 2011.
40. A. Touil, S. Chemkhi, and F. Zagrouba, “Moisture diffusivity and shrinkage of fruit and cladode of Opuntia ficus-indica during infrared drying,” *J. Food Process.*, vol. 2014, pp. 01–09, 2014.
41. İ. Doymaz, “Drying behaviour of green beans,” *J. Food Eng.*, vol. 69, no. 2, pp. 161–165, 2005.
42. M. Zielinska and M. Markowski, “Air drying characteristics and moisture diffusivity of carrots,” *Chem. Eng. Process. Process Intensif.*, vol. 49, no. 2, pp. 212–218, 2010.
43. M. I. H. Khan, C. Kumar, M. U. H. Joardder, and M. A. Karim, “Determination of appropriate effective diffusivity for different food materials,” *Dry. Technol.*, vol. 35, no. 3, pp. 335–346, 2017.
44. M. U. H. Joardder, C. Kumar, and M. A. Karim, “Multiphase transfer model for intermittent microwave-convective drying of food: Considering shrinkage and pore evolution,” *Int. J. Multiph. Flow*, vol. 95, pp. 101–119, 2017.

4 Numerical Model Formulation and Solution Approaches

4.1 INTRODUCTION

Mathematical models of drying need a theoretical basis from the perspective of physics, biology, chemistry, and material science, whereas the solution of the models needs the concepts of mathematics and computer science. Using theoretical basis and analytical tools, the transformation of phenomenological information to quantitative information can be accomplished. Even close prediction of quantitative information from available phenomenological data is possible with the advent of modern analytical tools and computation software. Several motivations regarding the numerical solution of drying phenomena can be proposed including the following ones:

- **Analytical solution** of a physical problem often deals with the solution of PDEs considering the boundary conditions. It provides mainly point solution of the required parameters, including temperature and moisture, at a given time. However, practical problems and their solution require overall temporal and spatial distribution of different variables. For example, the distribution of moisture and temperature is important in designing an appropriate drying system [1].
- **Practical problems** involve complicated irregular domain, complex boundary conditions, and a remarkable number of parameters. The analytical solution cannot deal with a problem with these constraints. In this case, the numerical solution using modern computation tools can provide an acceptable approximate solution to such complex problems. Recent advancement of computational capacity is now accessible due to the availability of high-performance computers and software platforms.
- **Due to the complex nature** of real-life problems, a parametric analysis is essential for checking the optimum conditions for designing a drying system. Parametric analysis is a great advantage of the numerical solution. A wide range of observation values can be attained from a base model by incorporating minor modification. This type of parametric analysis is not possible using an analytical solution.
- **Although analytical solutions provide the exact solution**, the formulations of the physical problem are at the basic level, considering many

DOI: 10.1201/9780429461040-4

simplifications. Therefore, this solution approach is not compatible with real-life problems. However, the numerical analysis provides the relatively approximate solution considering the real scenario of real-life problems.

- **Drying involves simultaneous heat and mass transfer** along with significant deformation. The analytical solution is not capable enough to deal with a problem that involves multiphysics, whereas the numerical solution offers an efficient solution for such coupled multiphysics problems.

Taking all these into consideration, a numerical solution is a perfect tool for solving real-life problem. However, an in-depth understanding of the underlying physics of drying and its analytical solution needs to be mastered first. A better understanding of the analytical solution can lead towards accomplishing numerical solutions of real-life problems. In the following section, we will present different strategies of modelling drying phenomena and their solution approach. For consistency in the discussion, we will use the examples of food materials from this chapter and onwards.

4.2 TYPES OF MATHEMATICAL MODELLING OF DRYING

Different modelling approaches are used for understanding and predicting drying kinetics and their associated variables. Drying models can range from empirical to theoretical depending on the accuracy and computation cost. The individual model has its advantages as well as shortcomings, as demonstrated in Figure 4.1. In brief, empirical models are observation-based, whereas theoretical models are completely physics-based.

Further classification of theoretical models can be made. During drying, phase change takes place. Simplified theoretical models consider only water migration as its gaseous form, and this type of model is referred to as a single-phase or diffusion model. The more comprehensive model considers transfer mechanisms of

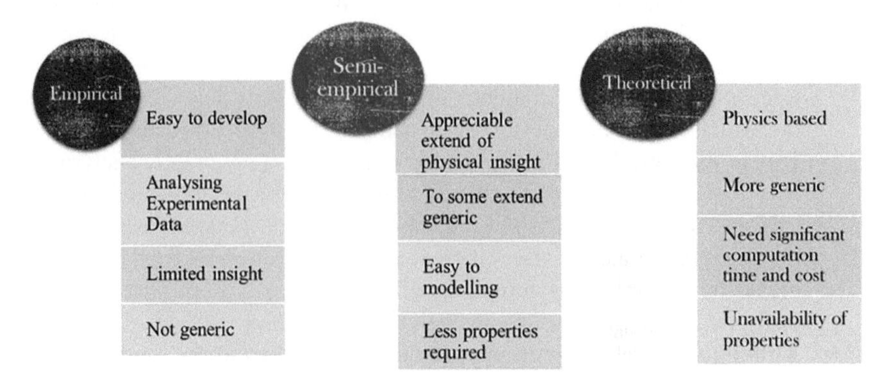

FIGURE 4.1 Drying models and their characteristics.

both liquid and gaseous phases of water. This type of model is generally referred to as a multiphase model. Moreover, depending on the convective heat and mass transfer constants and fluid flow, the model can be further subcategorized. Considering fluid flow, the model can provide spatially and temporally variable convective boundary conditions of the models [2,3].

Furthermore, deformation cannot be ignored on many occasions of drying. Incorporation of deformation causes enrichment of the drying model to go a step ahead [4]. Drying phenomena can be treated as a multiscale phenomena. Based on length scale, drying models can be classified as a micro-scale model and macroscale model. Combining the advantages of micro- and macro-scale modelling, a relatively new modelling paradigm called multiscale modelling can also be used for drying phenomena. In Chapters 5–11, different models will be discussed in detail, and the outline of the models is presented in Figure 4.2.

4.2.1 Empirical Modelling

Empirical models are completely experimental observation-based. This type of model is used widely as a mathematical tool to design drying systems in most drying fields. Formulation of the empirical model is simple enough that it does not need any prior expertise in the field of drying. Moreover, understanding of

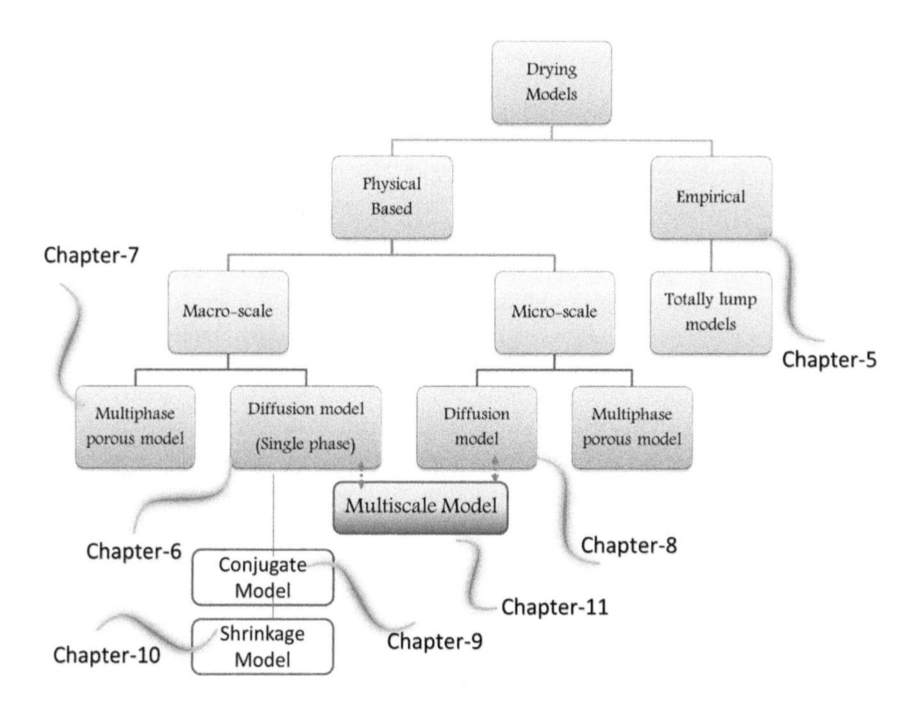

FIGURE 4.2 Outline of different drying models provided in this book.

these types models is possible with little training. Generally, these models use experimental data to develop the relationship between the moisture content of the sample and drying time. The fitting parameters used in the empirical equation of these models do not carry any physical significance.

Development of generic empirical modelling of drying kinetics that can be fit for different materials is not possible as these empirical relationships significantly change with the slight change of drying conditions and sample materials. In other words, these models are not flexible due to their dependency on the drying conditions and material properties.

4.2.2 Single-Phase Modelling

A single-phase model is the simplest form among the theoretical models. This model is based on the popular Fick's law of diffusion. In this model, the whole sample is treated as an object of a single phase. This model provides a preliminary theoretical understanding of temperature and moisture distribution in the sample during drying. As this considers only diffusion mass transfer mechanism and disregards other forms of internal mass transfer mechanisms, deep insight into important phenomena such as pressure-driven flow and liquid–gas phase change cannot be achieved. The prediction of temperature and moisture distribution is erroneous as a result of neglecting other mass transfer phenomena of multiphase species.

4.2.3 Multiphase Modelling

Drying is a simultaneous heat and mass transfer, facilitating the transfer of multiphase species from the porous sample. The migration of water with both liquid and gaseous phases is considered in the multiphase model [5]. In this modelling approach, the sample can be treated as hygroscopic porous material, which is a more realistic assumption for most of the materials that are subjected to drying [6,7]. All three phases, namely, solid, liquid, and gases, exist combined in a domain in a multiphase porous model [8]. Consideration of hygroscopic domain along with three species allows the model to consider other internal moisture transfer mechanisms including capillary flow, liquid diffusion, and so on.

4.2.4 Micro-Scale Modelling

Most of the physical phenomena, including drying, can be treated at different length scales as well as time scales [9]. Temperature and moisture distribution vary significantly at the micro-level from its macro-scale counterpart. Micro-level changes during drying are critical for developing an optimum drying process. Micro-structural characteristics of the material are the determinant physical properties that profoundly affect heat and mass transfer during drying. Micro-scale modelling offers many advantages including heterogeneous moisture and temperature distribution within the sample. One of the challenges in micro-scale modelling is real-time micro-level geometry [10,11]. Moreover, micro-level modelling provides too much detailed information at the smaller scale of the material, and is sometimes difficult to process at the scale of observation, namely at the macro-scale. Moreover, the micro-scale model requires high computational time.

4.2.5 Conjugated Drying Models

Drying involves simultaneous heat and mass transfer process that is significantly affected by the drying conditions. Temperature, airflow, humidity, and pressure are among the common parameters of drying [12]. These factors affect the boundary condition of the sample; especially, the conventional hot-air drying is affected most due to the variation of these parameters. Taking these into consideration, conjugation of fluid flow in simultaneous heat and mass transfer model makes the drying model more realistic [13]. This type of model simulates the temperature and moisture distribution for both the sample and surrounding of the sample. Moreover, the distribution of air velocity and pressure can also be obtained from the conjugate model. Eventually, the coupling of fluid flow, heat transfer, and mass transfer can offer a great tool for optimizing the drying condition for an efficient dryer.

4.2.6 Drying Model Considering Deformation

Deformation is an indispensable physical phenomenon observed in materials during different drying processes. Due to its heterogeneous hierarchical structure of porous materials, including food, many of the materials are highly shrinkable while drying [14]. Most importantly, shrinkage is an important factor that significantly affects the drying rate as well as drying kinetics.

Taking these concerns into consideration, deformation should not be neglected while predicting actual heat and mass transfer during drying. Several researchers proposed mathematical expressions to predict the deformation of food materials as a function of the moisture content. These models can be grouped into two categories: (i) theoretical models that are built based on the understanding of the fundamental physics and mechanisms that may be involved in deformation [15–17], and (ii) empirical models that are built by fitting the model's parameters to the experimental data [18–20].

Solid matrix of the porous materials shows stiffness of different degree. This solid matrix encounters noticeable stress due to water migration. The internal stress developed by the moisture gradient is the main facilitator of deformation during drying. Besides this, thermal stress causes deformation of the solid matrix in compensating the water migration. Moreover, pore formation is also a common phenomenon that can be observed during drying.

Combination of shrinkage and pore formation needs to be taken into consideration in the deformation model during drying. The model that considers deformation can predict the more accurate temperature and moisture distribution in the real-time domain than the model that deals with a non-deformable domain.

4.2.7 Multiscale Modelling

Most of the materials and physical phenomena can be observed as multiscale in nature. Magnifying of materials allows observing different characteristics of the material not observable at their visible range. Similar to this, slowing down or quickening time lapse provides more information than that at the normal speed [21]. Macroscale model uses average thermo-physical properties in the governing physics that take place at this scale. However, macrolevel transport mainly depends on micro-level transport phenomena and microstructure of the sample. Therefore, micro-level-derived transport properties need to be incorporated in the macro-scale model as they can provide realistic drying kinetics as well as the real distribution of temperature and moisture within the sample during drying. Multiscale offers this facility of bridging among the models of different temporal and spatial scales.

4.3 SOLUTION METHODS

Most of the physical phenomena that take place in nature are nonlinear in nature. These phenomena can be expressed with established laws with appropriate assumptions and simplifications. Then, the physical phenomena can be expressed in the language of science (mathematics). Due to the involvement of multivariable, namely, space and time, most of the physical phenomena associated conservation of mass, momentum, and energy can be expressed using partial differential equations (PDEs). This form of expression can able to express the phenomena as a continuous approach. To attain the solution of PDEs, both analytical and approximate solutions are available as shown in Figure 4.3.

As discussed earlier in this chapter, approximate solutions using numerical analysis is an acceptable option. As the exact solutions of PDEs involve an infinite series of variables, an analytical solution is time-consuming and inconvenient. There are many approaches to numerical solutions available in continuum and discrete approaches. However, finite element, finite volume, and finite difference methods are the most common solution methods for continuous physical phenomena. The main theme of these methods is to transfer continuous PDEs into a set of algebraic relations. Different considerations dealing with domain and transformation methods of the differential equations into algebraic equations make each method different from another.

4.3.1 Finite Element Method (FEM)

A system can be divided into an infinite number of pieces if it is considered as a continuum. The number of degree of freedom (DOF) is also infinite in this case. However, FEM divides the domain into finite small elements, and the resultant DOF is also deduced to finite. Among the advantages of FEM are its handling of complex geometry and the presence of a wide variety of engineering problems capable of being solved [22].

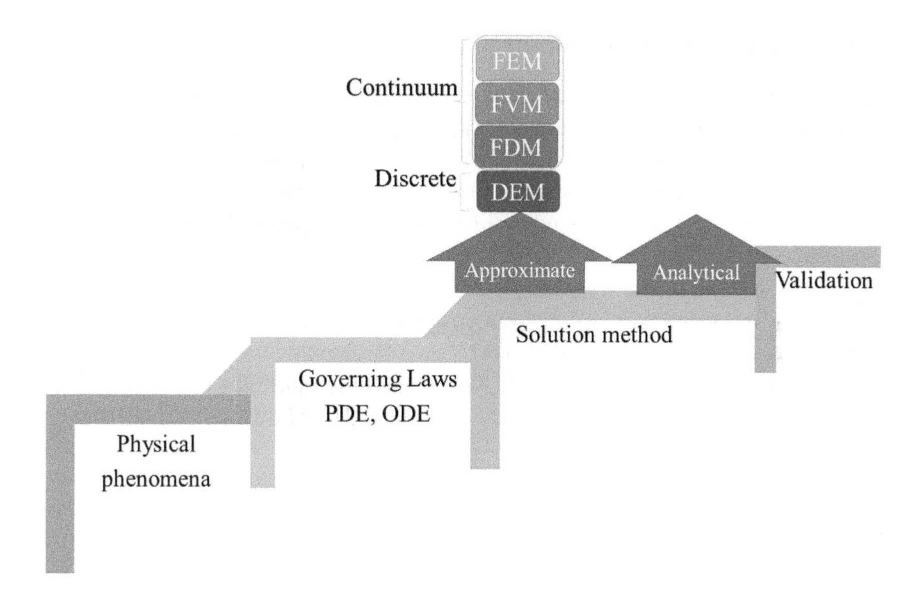

FIGURE 4.3 Solution steps of physical phenomena using numerical solutions.

Moreover, FEM can deal with multiphysics analysis, which is an important concern in many unit operations including drying. The problems related to structural analysis, transport phenomena including heat and mass transfer, fluid dynamics, and multiphysics can be solved using FEM.

Meshing is conducted using space discretization, and the elements are connected with sheared nodes of the elements. The element size (mesh) and shape are crucial in observing the distribution of temperature and moisture within the sample. Too large or too small of an element size sometimes leads to omitting the significant effect of certain physics at a different level. Mesh convergence or grid independency needs to be assessed carefully prior to computing the numerical problems using one of the continuum methods FEM, FDM, or FVM. The purpose of mesh convergence assessment is to reduce computational cost along with ensuring high accuracy of the solution.

After appropriate meshing of the domain, simplification of the complex PDEs of the elements needs to be executed. Using the Galerkin method, the approximation of the PDEs is generally accomplished in FEM. After the simplification, the steady problem can be represented by a set of algebraic equations; whereas, a set of ODEs can represent transient problems.

The simplified element equations are then solved using numerical linear algebra (NLA) method for the steady-state problem. In contrast to this, numerical integration techniques such as Runge-Kutta or Euler's methods are used to solve sets of ODEs in transient problems. Finally, the solutions of local (elemental) system are used to generate global (whole domain) systems of equations [22].

4.3.2 Finite Volume Method (FVM)

The finite volume deals with the natural phenomena and its associated complex mathematical expressions that are generally in PDEs. In FVM, a control volume is discretized into the finite number of smaller cells. One of the advantages of FVM over other continuum approaches is that it can deal with unstructured domain [23]. Therefore, the problems related to chemical engineering, fluid dynamics, and heat and mass transfer can be solved using FVM.

In this approach, the transfer phenomena are governed by the concepts of divergence of species, control volume, and boundary. Divergence of species (mass) across control volume and flux of species on the boundary layer are the main concepts of FVM. All of the PDEs are transformed in the divergence from using the divergent operator. The newly transformed divergence equations are integrated and the volume integral is converted into surface integral using Gauss's theorem. Eventually, the fluxes of variables across the domain boundary (control surface) of the cells can be obtained [23]. The subsequent steps, namely, attaining algebraic equation and solution of the sets of those equation follows a similar procedure as mentioned in FEM.

4.3.3 Finite Difference Method (FDM)

It is one of the simplest techniques of solving PDEs where a derivative of ODEs or PDEs is converted into the finite difference. In other words, the differential equation of a point of the system is converted to an algebraic equation. Therefore, the problems related to structural analysis and heat and mass transfer can be solved using FDM. Tailor series expansion is generally applied for the discretization of spatial domain and time interval. After that, matrix algebra techniques are applied to obtain the solution of the set of an algebraic equation. Although it is simple, it cannot be successfully applied in complicated and unstructured domains.

4.3.4 Discrete Element Methods (DEM)

Unlike continuum methods, discrete methods are one of the first principles of physics that considers the system as a domain consisting of numerous particles. The particle does not need to be the smallest unit of the sample; rather, a group of the smaller unit can be treated as a coarse-grain particle. The particles are not physically interconnected throughout the sample, whereas the finite components in all of the above-mentioned continuum methods are interconnected.

This method is advantageous for viscous and viscoelastic materials and frequently used to simulate hydrodynamic related problems [24]. The initial properties of the particles represent the initial state of the domain that evolves with time. Moreover, the particles can interact in response to the given forces applied during drying. DEM cannot provide the volume average information of continuum elements. A combined model of DEM and FEM/FVM can provide the corresponding deformation of a domain during drying.

Although the differences between discrete and continuum approach are quite clear, comparison of FEM, FVM, and FDM is not straightforward. The individual method offers distinguished advantages and encounters shortcomings of specific nature. One important concern is the ease of execution of the solution methods. Many basic aspects are similar in all above-mentioned methods. Therefore, a particular problem can be simulated using any one of these methods. However, selecting the most appropriate method needs a critical review of successful literature for a specific problem. Then, one should weigh the advantages of each solution method against potential shortcomings.

4.4 COMPUTATIONAL PLATFORMS AND VALIDATION

Once the governing physics of drying, properties of materials, drying conditions, scale of observation in length, and time are ready, simulation can be executed using computational software. The solving of numerical problems can be accomplished using user-developed code or commercial software package.

4.4.1 User-Developed Code

Users can develop their own codes by defining each of the parameters, boundary conditions, governing equations, and other related terms. User's own coding needs to start from scratch, and it is often the preferred pathways for research problems. Code can be written using programming languages including MATLAB, FORTRAN, VISUAL BASIC, and C++, among others.

Despite the user-controlled feature in user-developed code, the researcher encounters many problems in these tools. Enormous time consumption, difficulties in multiphysics coupling problems, and low reproducibility are the main challenges associated with user-developed code. To avoid these hurdles in user-developed code, commercial software is gaining popularity in the research community.

4.4.2 Computational Software

Commercial computational software have inbuilt features of defining material properties, developing geometry and meshing, and providing boundary conditions. Computational software is user-friendly at a different level depending on the complexity of the governing physics. Once material properties, the geometry of the sample, and boundary conditions are available to the researcher, the commercial software can do the rest of the things, as shown in Figure 4.4.

Identifying and solving unknown variables, selection of interpolation function and numerical solution-related operations are done on the software where no involvement of the user is required at all.

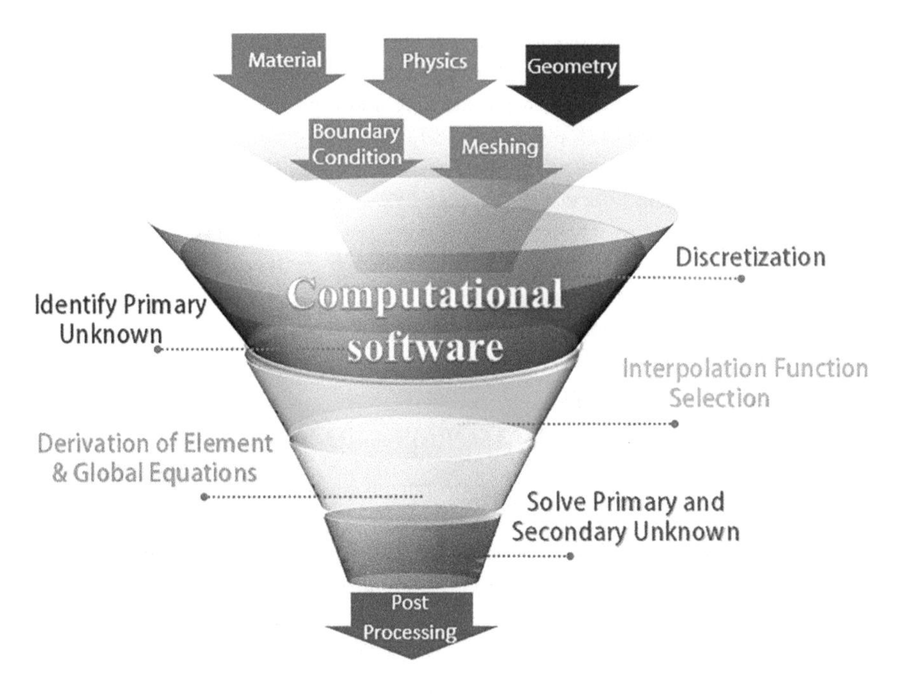

FIGURE 4.4 Numerical solution steps and the role of commercial software.

TABLE 4.1
Software Platform for Numerical Solutions

Numerical Solution Approach	Code/Software Package	Feature
FEM	FreeFEM, MFEM, ANSYS, COMSOL Multiphysics	All of the software offer inbuilt pre- and post-processing, meshing and geometry features. Apart from these, individual software offers different specialized features and continuous upgradation.
FVM	Flow 3D, OpenFOAM, StarCD	
FDM	FIDISOL	
DEM	LIGGGHTS, MercuryDPM	

There are numerous software platforms available for simulation of drying modelling as shown in Table 4.1, which is not intended to present the complete list of the computational software. Researchers and designers need to conduct a feasibility study before finding the most appropriate software for their specific modelling problems. All of the commercial software packages come with detailed guidelines in the form of documented materials as well as multimedia tutorials. The researcher needs to be familiar with the technical terms and work platforms of particular software.

Most of the recent commercial software offers a specific module for a particular physics. It is like a package/format of the computational modelling steps. Selecting an appropriate module (predefined physics) can be used as a base model and inserting the required values in the default model is sufficient for an elementary model solution. As a result of these advantages, less expertise in modelling is required for solving numerical solution than the user-developed code.

Multiphysics-related problems can be solved easily using commercial computational software. However, special attention is required to couple multiple physics in commercial software. Moreover, equation-based modelling is also available for the physical phenomena, and these are not readily available in the commercial software, which is still less time consuming than user-developed code. Despite this advantage, most of the popular commercial software it not free; rather, it is substantially expensive.

4.4.3 Validation of the Models

Mathematical modelling provides a deep insight into the underlying physics and assists in designing an appropriate system for unit operation including drying. Moreover, the modelling saves a huge amount of time and cost required for prototype fabrication. Further development and modification of the editing system can also be done using mathematical modelling. Despite these benefits of mathematical modelling, the model without validation cannot serve any role in engineering problem. Validation of a model is indispensable before applying its result in any engineering problems.

Simulated data of a numerical model needs to be done to check the accuracy of the model. The validation can be the authentication of all parameters or a couple of main parameters. Simulated results are often validated using experimental results or using data that are available in the literature. Moreover, approximated simulated results can be validated against benchmarked analytical solutions of the same governing equation. After getting both results of simulation and experimental, evaluation of the model, accuracy can be accomplished using statistical indicators. There are many statistical analysis options available, discussed in Chapter 5 in details.

Figure 4.5 demonstrates the validation of simulated data for moisture content during drying. The R^2 value indicates the goodness of the model data against the experimental data.

Validation should be done with the experimental data that maintains the same drying conditions and drying materials. The slight variation of experimental conditions or material properties of experiments from those in numerical modelling can make the data inaccurate for validation. Similarly, validation using data from the literature should not be done when there is an experimental arrangement option available.

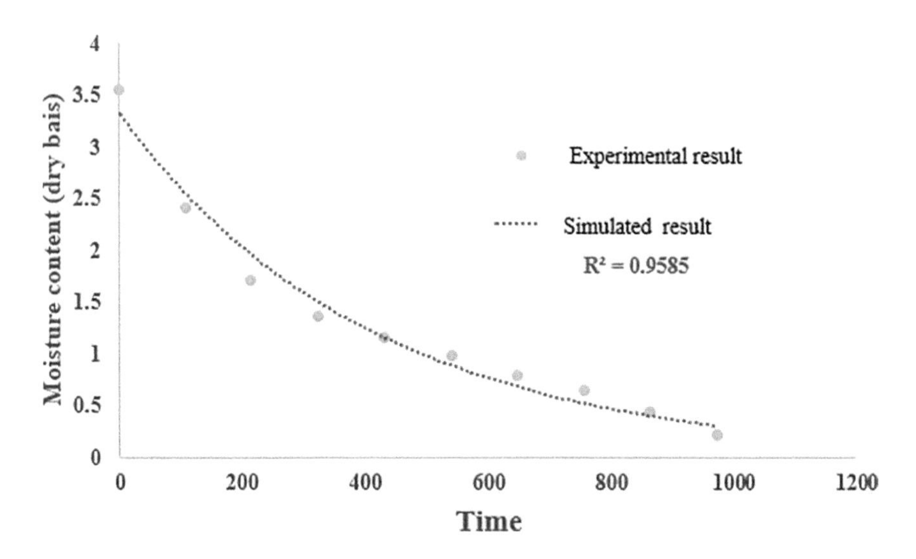

FIGURE 4.5 Simulated model validation using an experimental result.

REFERENCES

1. M. M. Rahman, M. U. Joardder, M. I. H. Khan, N. D. Pham, and M. A. Karim, "Multi-scale model of food drying: Current status and challenges," *Crit. Rev. Food Sci. Nutr.*, vol. 58, no. 5, pp. 858–876, 2018.
2. C. Kumar, M. U. H. Joardder, T. W. Farrell, and M. A. Karim, "Multiphase porous media model for intermittent microwave convective drying (IMCD) of food," *Int. J. Therm. Sci.*, vol. 104, pp. 304–314, 2016.
3. C. Kumar, M. U. H. Joardder, T. W. Farrell, G. J. Millar, and M. A. Karim, "Mathematical model for intermittent microwave convective drying of food materials," *Dry. Technol.*, vol. 34, no. 8, pp. 962–973, 2016.
4. M. U. H. Joardder, R. J. Brown, C. Kumar, and M. A. Karim, "Effect of cell wall properties on porosity and shrinkage of dried apple," *Int. J. Food Prop.*, vol. 18, no. 10, pp. 2327–2337, 2015.
5. C. Kumar, M. U. H. Joardder, T. W. Farrell, and M. A. Karim, "Investigation of intermittent microwave convective drying (IMCD) of food materials by a coupled 3D electromagnetics and multiphase model," *Dry. Technol.*, vol. 36, no. 6, pp. 736–750, 2018.
6. M. U. H. Joardder, C. Kumar, and M. A. Karim, "Multiphase transfer model for intermittent microwave-convective drying of food: Considering shrinkage and pore evolution," *Int. J. Multiph. Flow*, vol. 95, pp. 101–119, 2017.
7. M. U. H. Joardder, C. Kumar, and M. A. Karim, "Prediction of porosity of food materials during drying: Current challenges and directions," *Crit. Rev. Food Sci. Nutr.*, vol. 58, no. 17, pp. 2896–2907, 2018.
8. C. Kumar, M. U. H. Joardder, T. W. Farrell, G. J. Millar, and A. Karim, "A porous media transport model for apple drying," *Biosyst. Eng.*, vol. 176, pp. 12–25, 2018.
9. M. U. H. Joardder, C. Kumar, R. J. Brown, and M. A. Karim, "A micro-level investigation of the solid displacement method for porosity determination of dried food," *J. Food Eng.*, vol. 166, pp. 156–164, 2015.

10. M. I. H. Khan, M. U. H. Joardder, C. Kumar, and M. A. Karim, "Multiphase porous media modelling: A novel approach to predicting food processing performance," *Crit. Rev. Food Sci. Nutr.*, vol. 58, no. 4, pp. 528–546, 2018.
11. M. U. H. Joardder, C. Kumar, and M. A. Karim, "Food structure: Its formation and relationships with other properties," *Crit. Rev. Food Sci. Nutr.*, vol. 57, no. 6, pp. 1190–1205, 2017.
12. M. I. H. Khan, C. Kumar, M. U. H. Joardder, and M. A. Karim, "Determination of appropriate effective diffusivity for different food materials," *Dry. Technol.*, vol. 35, no. 3, pp. 335–346, 2017.
13. M. H. Masud, T. Islam, M. U. H. Joardder, A. A. Ananno, and P. Dabnichki, "CFD analysis of a tube-in-tube heat exchanger to recover waste heat for food drying," *Int. J. Energy Water Resour.*, vol. 3, no. 3, pp. 169–186, 2019.
14. M. H. Masud, M. U. H. Joardder, and M. A. Karim, "Effect of hysteresis phenomena of cellular plant-based food materials on convection drying kinetics," *Dry. Technol.*, vol. 37, no. 10, pp. 1313–1320, 2019.
15. J. E. Lozano, E. Rotstein, and M. J. Urbicain, "Shrinkage, porosity and bulk density of foodstuffs at changing moisture contents," *J. Food Sci.*, vol. 48, no. 5, pp. 1497–1502, 1983.
16. N. P. Zogzas, Z. B. Maroulis, and D. Marinos-Kouris, "Densities, shrinkage and porosity of some vegetables during air drying," *Dry. Technol.*, vol. 12, no. 7, pp. 1653–1666, 1994.
17. M. Mahiuddin, M. I. H. Khan, N. Duc Pham, and M. A. Karim, "Development of fractional viscoelastic model for characterizing viscoelastic properties of food material during drying," *Food Biosci.*, vol. 23, pp. 45–53, 2018
18. J. E. Lozano, E. Rotstein, and M. J. Urbicain, "Total porosity and open-pore porosity in the drying of fruits," *J. Food Sci.*, vol. 45, no. 5, pp. 1403–1407, 1980.
19. M. G. R. Perez and A. Calvelo, "Modeling the thermal conductivity of cooked meat," *J. Food Sci.*, vol. 49, no. 1, pp. 152–156, 1984.
20. R. S. Rapusas and R. H. Driscoll, "Thermophysical properties of fresh and dried white onion slices," *J. Food Eng.*, vol. 24, no. 2, pp. 149–164, 1995.
21. M. M. Rahman, M. U. H. Joardder, M. I. H. Khan, N. D. Pham, and M. A. Karim, "Multi-scale model of food drying: Current status and challenges," *Crit. Rev. Food Sci. Nutr.*, vol. 58, no. 5, pp. 858–876, 2018.
22. J. N. Reddy, *An Introduction to the Finite Element Method*, vol. 1221. McGraw-Hill New York, 2004.
23. O. Kolditz, *Computational Methods in Environmental Fluid Mechanics.* Springer Science & Business Media, 2002.
24. H. C. P. Karunasena, R. J. Brown, Y. Gu, and W. Senadeera, "Application of meshfree methods to numerically simulate microscale deformations of different plant food materials during drying," *J. Food Eng.*, vol. 146, pp. 209–226, 2015.

5 Empirical Modelling of Drying

5.1 INTRODUCTION

The empirical model is the most widely used mathematical tool to predict the drying kinetics in the various fields of drying. Most of the drying industries still use various empirical modelling techniques to design drying systems. The main reason for its prevalence is its simplicity in use, which does not need any prior expertise in the field of drying. Empirical models are developed using experimental data to fit parameters for a particular system. Methods, materials, and the processing environment restrict these models, and the fitting parameters usually have no physical significance. Even with these limitations, the empirical models still provide a good prediction. Despite offering this convenience of implementation in the drying process, empirical modelling possesses some critical shortcomings, which will be discussed at the end of this chapter.

5.2 REGRESSION ANALYSIS

Regression analysis is a statistical tool utilized for the development of a relationship between dependent variables and independent variables. Independent variables can be more than one in many scenarios. The obtained relationships can be used as prediction tools for future use of the same unit process. Based on the nature of trends of the relationships between dependent variable and independent variable, the regression analysis can be categorized as simple linear, multiple linear, and nonlinear regression. Choosing of the type of regression analysis depends on the nature of data sets. Linear regression is the most common regression analysis, while nonlinear approach is only used in order to deal with complicated data sets.

5.2.1 Simple Linear Regression

If there are only two variables and the regression curve appears straight, linear regression is a good choice. The simple linear regression is generally expressed with the following equation (Figure 5.1):

DOI: 10.1201/9780429461040-5

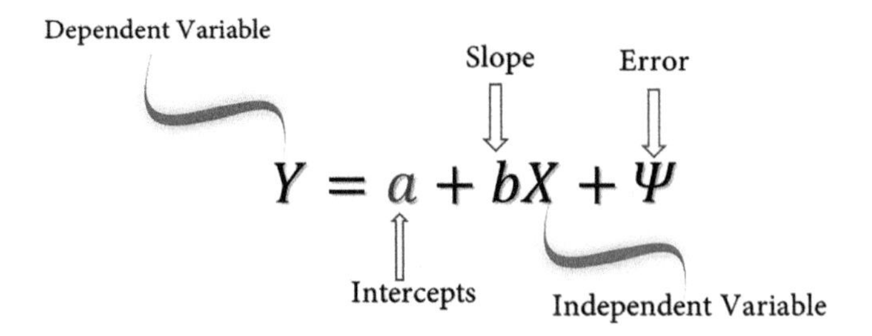

FIGURE 5.1 Simple linear regression.

5.2.2 Multiple Linear Regression

If the dependent variables are more than one, multiple linear regression can be fitted in these cases. If $X_1, X_2, \ldots, X_n$ are the dependent variables, then the mathematical expression of multiple linear regression is formed as follows:

$$Y = a + bX_1 + cX_2 + \cdots + nX_n + \Psi \tag{5.1}$$

In the case of multi-variable linear regression, non-collinearity, the minimum correlation among independent variables, is taken into consideration.

Although linear regression generally deals with straight-line relationships among the variables, linear regression can incorporate polynomial terms, reciprocal terms, and log function. In other words, linear models can contain square, log, and inverse terms in it to fit a curve.

5.2.3 Non-Linear Regression

For most of the natural phenomena, the linear regression model cannot be used as it does not fit with the experimental data. In this case, nonlinear regression is an alternative option. It is not the nature of the line, whether it is straight or curved, that determines the selection of linear or nonlinear model; rather, nonlinear regression is a straightforward choice for the trends of variables that cannot be fitted in linear regression. One of the main features of nonlinear regression is the enormous flexibility of the model. In other words, nonlinear regression offers several forms that can fit the curves or trends of the variables. For instance, power, Weibull growth, and Fourier are some of the common nonlinear regression analyses. Due to the vast number of options, it is sometimes difficult to choose the right nonlinear function to fit the curve.

Although there are various methods to fit a curve, identifying the best model needs considerable effort. Following are some of the points that need to be taken into consideration before developing an empirical model:

- Knowledge of drying of particular materials and the standard practice of the field often can assist
- Adequate amount of data and right approach of data acquisition

- Choosing the minimum number of independent variables and keeping the important ones
- Development of the model as simple as possible

After the general discussion relating to regressing analysis which is the backbone of developing an empirical model, we will move to a detailed discussion on empirical modelling pertaining to drying of materials.

5.3 EMPIRICAL MODELLING FOR THE DRYING PROCESS

Empirical modelling is used as an important tool for designing drying systems, attaining optimization of the existing drying systems, as well as establishing the nature of kinetics of transport phenomena and quality attributes. From the empirical model, one can also predict the energy consumption of the drying process. In general, the empirical model is commonly used for developing drying kinetics trends for a specific product. Despite these benefits of empirical models, development of a generic model that fits for all materials and the drying condition is impossible.

Types of drying, conditions of drying (including temperature, air velocity, and relative humidity), material properties, and even sample geometry significantly affect the drying kinetics models. However, all of these factors do not influence equally, for instance, drying temperature and sample thickness are the crucial variables that affect the most in the empirical model for a sample in conventional hot-air drying [1].

5.3.1 Important Considerations of the Empirical Model

- **In general, most of the empirical models of drying** are developed for predicting moisture content in the sample with drying time. However, there are other empirical models available that deal with other parameters and quality. Although this chapter deals with drying kinetics models, a short discussion of the empirical model is presented in connection with the quality prediction model at the end of this chapter.
- **Consideration of drying conditions:** Most of the drying kinetics models deal with moisture content as a dependent variable and time as an independent variable. There are very few empirical drying kinetics models that incorporate other parameters other than time to predict moisture content.
- **The trend of model:** As discussed in the previous section, different types of trends can be modelled using an empirical modelling approach. The most common drying models use simple linear regression analysis. However, a few models of nonlinear nature are also available in predicting drying kinetics.
- **Several fitting constants:** To attain good fit models, several fitting constants are required. Less number of fitting parameters are expected for a convenient model as this reduces the number of data required for the prediction model.

5.3.2 Drying Kinetics Models

The process of drying commonly means moisture evaporation due to simultaneous heat and mass transfer. Typical drying kinetics is shown in Figure 5.2. The surface contains free moisture, and therefore, during the first stage of drying, a constant drying rate is observed. In this stage, vaporization of this free moisture takes place. The drying rate at this stage is surface-based and governed by external factors such as the area exposed to the dry air, the temperature difference between wet surface and dry air, as well as the external heat and mass transfer coefficients.

After this period, to facilitate drying, the moisture has to be diffused from the inside of the material to the surface. Therefore, beyond the point called critical moisture content (M_c), the drying rate moves from the constant rate stage to first falling stage. In this phase, the drying rate is mainly dependent on thickness, shape, and the collapse of internal tissues. Heat damage may occur at the surface at this stage due to lack of moisture. In the first falling-rate period, the drying rate reduces as the moisture content decreases because of the additional internal resistance for moisture transfer and the reduction of heat flux.

In the second falling rate, moisture moves from the centre to the surface due to diffusion, resulting from the concentration gradient between the core region and the surface. The diffusion rate decreases in the falling rate drying period due to shrinkage and lowers the moisture gradient that results in a longer drying time. The drying rate of this later falling rate period is slow. To remove the last 10% of the water from materials it takes almost equal time to that required for removing the first 90% of the water content.

Drying behaviour of some biological and most food materials experience this second falling rate period. Supplying more energy by increasing the drying air

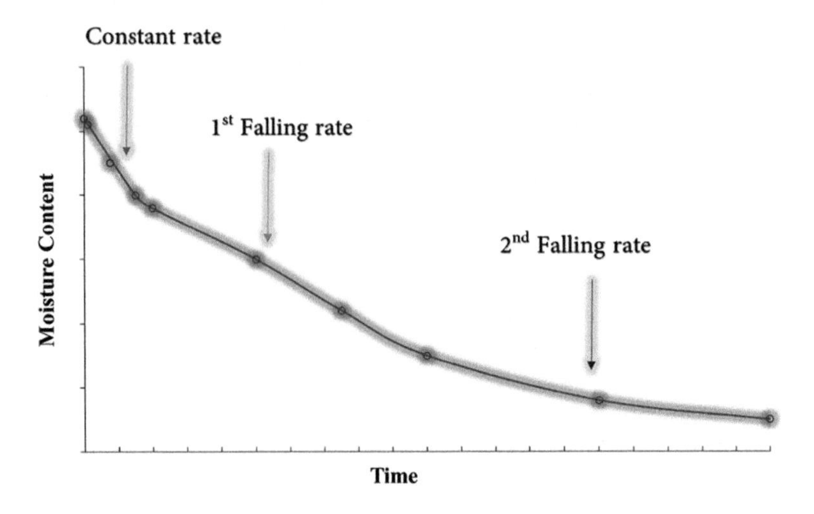

FIGURE 5.2 Typical drying curve showing constant and falling rates.

temperature in this stage accelerates drying. But higher temperature may damage the surface of the product, resulting in hard crust formation. The moisture content at the centre is still high because most of the porous materials are poor heat conductors, and therefore limit the supply of heat from the surface to the inner part. Thus, the surface becomes overheated and damaged. Table 5.1 summarizes the applications of different drying models for different food materials.

TABLE 5.1
Application of Empirical Models in Different Drying and Samples [2–50]

	Best Fit Condition and Materials		
Model	**Materials**	**Drying Condition**	**Remarks**
Midilli and others	Apple	TD, HCD	CTD: Cabinet tray dryer OD: Oven dryer TCD: Tunnel convection dryer HCD: Hot-air convective dryer MWD: Microwave dryer IRD: Infrared dryer STD: Solar tunnel dryer SD: Sun drying ISD: Indirect solar dryer FBD: Fluidized bed dryer OMD: Osmotic dehydration
	Chilli	OD and FBD	
	Mango	MWD, HCD	
	Pineapple	IRD+HCD	
	Pumpkin slices	TD, HCD	
	Saffron	IRD	
Logarithmic	Apple	HCD	
	Basil leaves	TTD	
	Pumpkin	CTD	
Diamante and others	Apricot	HCD	
Peleg	Banana	HCD, OMD	
Wang and Singh	Banana	MWD, ISD	
Page	Bitter melon	HCD	
	Date palm fruit	MWD	
	Green bean	SD	
	Kiwifruit	CTD	
Modified page (II)	Basil leaves	SD	
Modified page (III)	Onion slices	IRD+HCD	
Thompson	Blueberries	IRD	
	Green peas	HCD	

(Continued)

TABLE 5.1 (*Continued*)
Application of Empirical Models in Different Drying and Sample [2–50]

	Best Fit Condition and Materials		
Model	**Materials**	**Drying Condition**	**Remarks**
Aghbashlo and others	Carrot	HCD	
Verma and others	Carrot slices	IFD	
	Parsley	SD	
Hii and others	Carrot	HCD	
Two-term	Fig	TTD	
	Onion slices	SD	
	Plum	HCD	
Weibull	Garlic	HCD	
Approximation of diffusion	Green pepper	SD	
Demir and others	Green table olives	HCD	
Modified Midilli and others	Jackfruit	HCD	
Newton	Red chilli	HCD	
	Strawberry	SD	
Two-term exponential	Starfruit slices	TD	
Midilli and others;	Apple slices	OD	
Page; logarithmic	Banana slices	HCDF	
Two-term; logarithmic	Beetroot	TD and MD	
Modified Henderson and Pabis	Pumpkin	HCD	
Logarithmic; Verma and others	Pumpkin slices	CTD	
Two-term; logarithmic	Pumpkin	HCD	

The table presents some selected empirical models of different materials. There are numerous such occasions where empirical relations are used to predict drying kinetics and dryer design. The mentioned models can be subcategorized as empirical and semi-empirical, as discussed below.

5.3.3 Empirical Models

Empirical models refer to a direct relationship between the average moisture content and drying time. These types of models are derived from experimental data and usually lack comprehensive physical interpretation [51]. The most suitable models that adequately describe the drying kinetics of some fruits and vegetables are shown in Table 5.2.

TABLE 5.2
Summary of Empirical Relationships of Different Models

Model	Empirical Relation	Involved Function	No. of Terms	Fitting Parameters	Ref.
Peleg	$MR = \frac{(M - M_e)}{(M_0 - M_e)} = 1 - t/(a + bt)$	1st degree	02	a, b	[52]
Wang and Singh	$MR = \frac{(M - M_e)}{(M_0 - M_e)} = 1 + at + bt^2$	2nd degree	03	a, b	[14]
Diamante and others	$\ln(-\ln MR) = a + b(\ln t) + c(\ln t)^2$	Logarithmic	03	a, b, c	[10]
Thompson model	$t = a\ln(MR) + b[\ln(MR)]^2$	Logarithmic	02	a, b	[53]
Aghbashlo and others	$MR = \frac{(M - M_e)}{(M_0 - M_e)} = \exp\left(\frac{K_1 t}{1 + K_2 t}\right)$	Exponential	01	K_1, K_2	[24]
Weibull model	$MR = \frac{(M - M_e)}{(M_0 - M_e)} = \infty - b\exp(-k_0 t^n)$	Exponential	02	∞, b, k_0	[33]
Silva and others	$MR = \frac{(M - M_e)}{(M_0 - M_e)} = \exp(-at - b\sqrt{t})$	Exponential	01	a, b	[54]

5.3.4 Semi-Empirical Models

The semi-empirical models are not completely experimental data fitting based, although parameter-based models do have some physical insight. These models also depend on the experimental results for some particular parameters. Semi-empirical models are mainly based on Newton's law of cooling and Fick's law of diffusion. Examples of semi-empirical models which are commonly employed in predicting drying kinetics are briefly discussed below.

5.3.4.1 Models Derived from Newton's Law of Cooling

Semi-empirical models based on Newton's law of cooling are summarized in Table 5.3. These models use the exponential relationship of moisture ratio and drying time.

5.3.4.2 Fick's Law Based Semi-Empirical Models

A wide number of semi-empirical models have been developed based on Fick's second law of diffusion. Table 5.4 shows Fick's law-based semi-empirical drying models.

The fitting parameters depend on the sample properties and geometry as well as the drying conditions. They do not carry any physical signs at all.

5.4 QUALITY KINETICS MODEL

During the drying process, the distributions of temperature and moisture in the sample can characterize the degradation rate of the quality of dried foods. This distribution can be used as the continuous inputs in developing an analytical and kinetic quality model to map the profiles of the food quality during drying. The change of quality index of dried products is described from the following kinetic order model:

$$\frac{-dA}{dt} = kA^n \tag{5.2}$$

where A is the quality index and n is the order of the reaction. The reaction rate constant, k, can be calculated from the Arrhenius law, given below (Equation 5.3), which depends on the temperature T of the sample during drying.

$$k = k_o e^{-\frac{\Delta E_a}{RT}} \tag{5.3}$$

where R is a universal gas constant, and ΔE_a and k_o are activation energy and pre-exponential factor, respectively, which can be calculated by a two-step linear or one-step nonlinear regression analysis of experimental data conducted at different drying temperatures.

It can be noted that the change of colour; structural changes such as deformation, shrinkage, porosity, and texture; change in nutritional content such as

TABLE 5.3
Summary of Newton's Law of Cooling-Based Empirical Relationships

Model	Empirical Relation	Involved Function	No. of Terms	Fitting Parameters	Ref.
Newton model	$MR = \frac{(M - M_e)}{(M_0 - M_e)} = \exp\left(-k^t\right)$	Exponential	01	k	[38]
Page model	$MR = \frac{(M - M_e)}{(M_0 - M_e)} = \exp\left(-kt^n\right)$	Exponential	01	k, n	[55]
Modified page (II)	$MR = \frac{(M - M_e)}{(M_0 - M_e)} = \exp\left(-(kt)^n\right)$	Exponential	01	k, n	[20]
Modified page (III)	$MR = \frac{(M - M_e)}{(M_0 - M_e)} = k\exp\left(-{t}/{d^2}\right)^n$	Exponential	01	k, d, n	[22]

TABLE 5.4
Summary of Fick's Law-Based Empirical Relationships of Different Models

Model	Empirical Relation	Involved Function	No. of Terms	Fitting Parameters	Ref.
Henderson and Pabis model	$MR = \frac{(M - M_e)}{(M_0 - M_e)} = a \exp(-kt)$	Exponential	01	a, k	[56]
Modified Henderson and Pabis model	$MR = \frac{(M - M_e)}{(M_0 - M_e)} = a \exp(-kt) + b \exp(-gt) + c \exp(-ht)$	Exponential	03	a, b, c, k, g, h	[57]
Midilli and others	$MR = a \exp(-kt) + bt$	Exponential	02	a, k, b	[2]
Logarithmic model	$MR = \frac{(M - M_e)}{(M_0 - M_e)} = a \exp(-kt) + c$	Exponential	02	a, k, c	[6]
Two-term	$MR = \frac{(M - M_e)}{(M_0 - M_e)} = a \exp(-K_1 t) + a \exp(-K_2 t)$	Exponential	02	a, b, K_1, K_2	[30]
Demir and others	$MR = \frac{(M - M_e)}{(M_0 - M_e)} = a \exp\left((-kt)^n\right) + b$	Exponential	02	a, b, k, n	[36]
Verma and others	$MR = \frac{(M - M_e)}{(M_0 - M_e)} = a \exp(-kt) + (1 - a)\exp(-gt)$	Exponential	02	a, k, g	[25]
Approximate diffusion model	$MR = \frac{(M - M_e)}{(M_0 - M_e)} = a \exp(-kt) + (1 - a)\exp(-kbt)$	Exponential	02	a, b, k	[34]

ascorbic acid and total polyphenol content can be predicted from empirical models. There are many empirical models for the mentioned quality individually. The interested reader is referred to the review article on the topic for detailed discussions.

5.5 VALIDATION AND INTERPRETING REGRESSION MODELS OUTPUT

To determine which relationships (empirical model) are significant and appropriate, the nature of the attained relationship, coefficients of regression, and *P*-value need to be calculated. Coefficients interpret the types of interrelationships that exist among the variables, whereas the *P*-value reflects the validity of the relationships. The importance of *P*-value and correlation coefficient has been presented in Figure 5.3.

To increase the reliability and reproducibility of the empirical models, all of the experiments need to be done in triplicate.

5.5.1 Regression Coefficients

Correlation coefficient denotes to what extend the independent variable affects the dependent variable. Both positive and negative correlation can be obtained for the coefficient. A positive value of the coefficient refers to the proportional effect of independent variables on the dependent variable, whereas the negative reflects the inverse relationships. Moreover, the value of coefficient denotes the level of accuracy of the model as well as the dependency of the dependent variables on the independent variable. There are various testing methods of the empirical model including the linear determination coefficient (R^2), root mean squared errors (RMSE), sum squared errors (r^2), and chi-square (χ^2).

In general, the higher the values of *coefficient*, the better the fit of the empirical model of drying of materials. The different empirical model is suitable for different materials and shows non-identical coefficients for individual samples and systems.

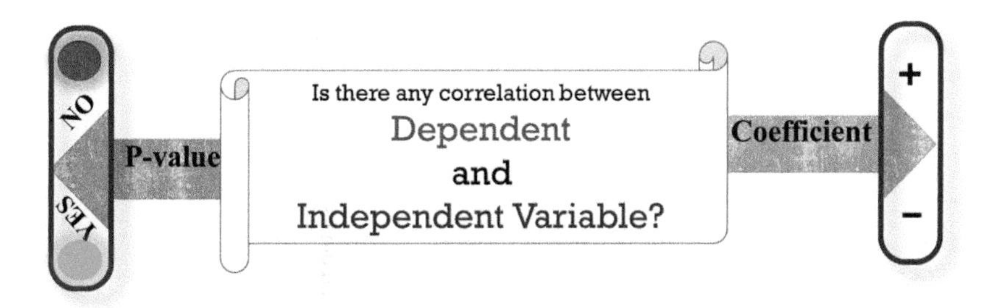

FIGURE 5.3 Validation and interpretation tools of the regression model.

5.5.1.1 *P*-Value

The *p*-values for the coefficients indicate whether the developed empirical relationships are statistically valid. The *p*-value of each independent variable assesses the effect of the concerned variable on the dependent variable. If there exists any correlation, there must be an association between the changes of the independent variable and deviation of dependent variables. In the usual significance level, the *p*-value is taken as 0.05 or below; any value more than this does not carry any statistical significance. Any variables that possess *p*-values greater than the standard value should be discarded; otherwise, it will reduce the precision of the model.

5.5.1.2 Chi-Square (χ^2)

The chi-square test compares the effect of two variables in a contingency table to observe whether they are interrelated. Chi-square value can be calculated from the following equation:

$$\chi_c^2 = \frac{\sum_{i=1}^{N} (Y_{ei} - Y_{pi})}{Y_{pi}} \tag{5.4}$$

pi and *ei* are the predicted and experimental values, respectively. The lower values for χ^2 approaching zero are considered optimal and can be a tool to choose the best model. The different empirical model shows different values of χ^2 for the same sample drying. This variation is caused due to the number of variables and terms of the empirical models. Similarly, χ^2varies for different samples for an individual model. For example, Modified Page model (I) was found to be the most suitable for describing the drying curve of mint and basil leaves with χ^2 of 0.000651254 and 0.00020655, respectively [58].

5.5.1.3 *R*-Squared

R-squared is the representation of the goodness of fit of a linear regression model. The higher values of R^2 confer a better fit of the empirical model of drying of materials; however, small R^2 is not necessarily bad for every situation [3,47,59]. The value of R^2 can be ranged from 0.-to1.00.The the empirical model that provide the R^2 value close to 1.00can be graded as a good the model. For instance, the best fit models of drying kinetics of different materials showed R^2=0.999 [57]. The value of R^2 is determined from the following relationship:

$$R^2 = \frac{\text{Variance explained by the model}}{\text{Total varience}} \tag{5.5}$$

$$R^2 = \frac{\sum_{i=1}^{n} \left(MR_i - MR_{pre,i}\right) \cdot \sum_{i=1}^{n} \left(MR_i - MR_{\exp,i}\right)}{\sqrt{\left[\sum_{i=1}^{n} \left(MR_i - MR_{pre,i}\right)^2\right] \cdot \left[\sum_{i=1}^{n} \left(MR_i - MR_{\exp,i}\right)^2\right]}} \tag{5.6}$$

where $MR_{exp,I}$ and $MR_{pre,i}$ are the *i*th experimentally observed and predicted moisture ratio, respectively. N is the number of observations, whereas n is the number of constants [58].

There is no single value of R^2 that reflects all of the best models. It varies with the variation of subject area as well as the number of associated variables. Sometimes, R^2 alone is not sufficient to determine whether the regression is biased or not as in some cases a biased model can provide high R^2 value. Over-fitting the regression model and inadequate data acquisition sometimes cause high R^2 in a deceptive good model. However, where drying modelling is concerned, the generally expected R^2 values are high.

5.5.1.4 Standard Deviation (SD)

The simplest way of calculating the deviation of a variable from the normal value is the standard deviation (SD). In other words, using the SD one can get that what is normal and what is the minimum and maximum deviation associated in the values obtained from the empirical model. From the SD, the standard error of the mean can also be calculated from the following equations:

$$\text{Standard deviation}, \sigma = \sqrt{\frac{\sum_{i=1}^{n}(x_i - \bar{x})}{n-1}} \tag{5.7}$$

$$\text{Standard error of the mean (SEM)}: \sigma_M = \frac{\sigma}{\sqrt{N}} \tag{5.8}$$

where x_i and $\bar{x}$ are the individual observation value and mean value, respectively. Here, n and N are the number of values and the number of observations, respectively.

5.5.1.5 Sum Square Error (SSE)

To assess the amount of errors involved in the predicted model, sum square error (SSE) is a handy tool. Similar to other errors, the lowest values of SSE close to zero are considered to be optimal. In drying, many empirical models offer low values of SSE, close to 0.001 [3,47,59]. SSE can be determined from the following expression:

$$SSE = \frac{1}{N}\sum_{i=1}^{N}(Y_{ei} - Y_{ei})_2 \tag{5.9}$$

5.5.1.6 Root Mean Square Error (RMSE)

Root mean square error (RMSE) is another good option to check the error in the predicted model. The lowest values for RMSE close to zero are considered to be optimal and can be used to choose the best prediction model. The RMSE value provides the deviation between the predicted and experimental values of a variable and can be calculated from the following expression [60]:

$$RMSE = \frac{\sqrt{\sum_{i=1}^{n}\left(MR_{i,pre} - MR_{i,\exp}\right)^{2}}}{N} \tag{5.10}$$

5.6 LIMITATIONS

Empirical models are based on fitting the model parameters to the experimental data. Even though the empirical models give a reasonably accurate prediction, they offer limited insight into the fundamental principles involved, limiting the understanding of the mechanisms responsible for change of quality. Apart from this, an empirical model has the following shortcomings:

- **Models are not generic:** An empirical model best fit for a sample does not necessarily fit well for the other materials. Moreover, a single model can only be used for the specific size and shape of the sample.
- **Specific drying types and conditions:** Applicability of empirical models also depends on the types of drying. Individual drying type results in unique drying conditions, eventually fitting diverse empirical models. Moreover, it also varies with the variation of drying conditions.
- **Adequate data acquisition:** Models must be developed using adequate numbers of experimental data. There are no specific rules for determining the number of adequate data. The required number of data varies with diverse factors including types of drying, drying conditions, sample materials, and sample compactness, among other factors.

Generic empirical modelling of drying kinetics materials is complex due to the involvement of many interrelated variables. A generic model can predict the drying kinetics of a product during drying regardless of the process. Although extensive research has been carried out on empirical model formulations, no single empirical model considers both process conditions and material properties. As a wider variety of processing conditions and various material properties need to be considered in developing the realistic and generic empirical model, it is therefore almost impossible in development.

REFERENCES

1. S. Azzouz, A. Guizani, W. Jomaa, and A. Belghith, "Moisture diffusivity and drying kinetic equation of convective drying of grapes," *J. Food Eng.*, vol. 55, no. 4, pp. 323–330, 2002.
2. A. Midilli, H. Kucuk, and Z. Yapar, "A new model for single-layer drying," *Dry. Technol.*, vol. 20, no. 7, pp. 1503–1513, 2002.

3. E. K. Akpinar, "Determination of suitable thin layer drying curve model for some vegetables and fruits," *J. Food Eng.*, vol. 73, no. 1, pp. 75–84, 2006.
4. W. A. M. McMinn, "Thin-layer modelling of the convective, microwave, microwave-convective and microwave-vacuum drying of lactose powder," *J. Food Eng.*, vol. 72, no. 2, pp. 113–123, 2006.
5. W. A. M. McMinn, C. M. McLoughlin, and T. R. A. Magee, "Thin-layer modeling of microwave, microwave-convective, and microwave-vacuum drying of pharmaceutical powders," *Dry. Technol.*, vol. 23, no. 3, pp. 513–532, 2005.
6. P. K. Chandra and R. P. Singh, Applied *Numerical Methods* for *Food* and *Agricultural Engineers*. CRC Press, 1995.
7. A. Kaleta and K. Górnicki, "Evaluation of drying models of apple (var. McIntosh) dried in a convective dryer," *Int. J. food Sci. Technol.*, vol. 45, no. 5, pp. 891–898, 2010.
8. O. Yaldiz, C. Ertekin, and H. I. Uzun, "Mathematical modeling of thin layer solar drying of sultana grapes," *Energy*, vol. 26, no. 5, pp. 457–465, 2001.
9. L. Diamante, M. Durand, G. P. Savage, and L. P. Vanhanen, "Effect of temperature on the drying characteristics, colour and ascorbic acid content of green and gold kiwifruits," *Intl. Food Res. J.*, vol. 451, pp. 441–451, 2010.
10. L. M. Diamante, R. Ihns, G. P. Savage, and L. Vanhanen, "A new mathematical model for thin layer drying of fruits," *Int. J. Food Sci. Technol.*, vol. 45, no. 9, pp. 1956–1962, 2010.
11. W. P. da Silva, A. F. Rodrigues, C. M. D. P. S. E. Silva, D. S. de Castro, and J. P. Gomes, "Comparison between continuous and intermittent drying of whole bananas using empirical and diffusion models to describe the processes," *J. Food Eng.*, vol. 166, pp. 230–236, 2015.
12. M. Ngoulou, R. G. Elenga, L. Ahouet, S. Bouyila, and S. Konda, "Modeling the drying kinetics of earth bricks stabilized with cassava flour gel and amylopectin," *Geomaterials*, vol. 9, no. 1, pp. 40–53, 2018.
13. N. Benmakhlouf, S. Azzouz, and A. El Cafsi, "Experimental and mathematical investigation of parameters drying of leather by hot air," *Heat Mass Transf.*, vol. 54, no. 12, pp. 3695–3705, 2018.
14. M. I. Fadhel, R. A. Abdo, B. F. Yousif, A. Zaharim, and K. Sopian, "Thin-layer drying characteristics of banana slices in a force convection indirect solar drying," in Proceedings of the 6th IASME/WSEAS International Conference on Energy and Environment (EE 2011), 2011, pp. 310–315.
15. S. Pusat and M. T. Akkoyunlu, "A new empirical correlation to model drying characteristics of low rank coals," *Int. J. Oil, Gas Coal Technol.*, vol. 15, no. 3, pp. 287–297, 2017.
16. J. Chen et al., "Mathematical modeling of hot air drying kinetics of *Momordica charantia* slices and its color change," *Adv. J. Food Sci. Technol.*, vol. 5, no. 9, pp. 1214–1219, 2013.
17. M. G. A. Vieira and S. C. S. Rocha, "Mathematical modeling of handmade recycled paper drying kinetics and sorption isotherms," *Brazilian J. Chem. Eng.*, vol. 25, no. 2, pp. 299–312, 2008.
18. N. B. Kardile, P. K. Nema, B. P. Kaur, and S. M. Thakre, "Comparative semi-empirical modeling and physico-functional analysis of hot-air and vacuum dried puran powder," *J. Food Process Eng.*, vol. 43, no. 1, p. e13137, 2020.
19. D. Sridhar and G. M. Madhu, "Drying kinetics and mathematical modeling of *Casuarina equisetifolia* wood chips at various temperatures," *Period. Polytech. Chem. Eng.*, vol. 59, no. 4, pp. 288–295, 2015.

20. G. M. White, I. J. Ross, and C. G. Poneleit, "Fully-exposed drying of popcorn," *Trans. ASAE*, vol. 24, no. 2, pp. 466–468, 1981.
21. L. M. Diamante and P. A. Munro, "Mathematical modelling of the thin layer solar drying of sweet potato slices," *Sol. Energy*, vol. 51, no. 4, pp. 271–276, 1993.
22. D. G. P. Kumar, H. U. Hebbar, and M. N. Ramesh, "Suitability of thin layer models for infrared–hot air-drying of onion slices," *LWT-Food Sci. Technol.*, vol. 39, no. 6, pp. 700–705, 2006.
23. J. Shi, Z. Pan, T. H. McHugh, D. Wood, E. Hirschberg, and D. Olson, "Drying and quality characteristics of fresh and sugar-infused blueberries dried with infrared radiation heating," *LWT-Food Sci. Technol.*, vol. 41, no. 10, pp. 1962–1972, 2008.
24. M. Aghbashlo, M. H. Kianmehr, S. Khani, and M. Ghasemi, "Mathematical modelling of thin-layer drying of carrot," *Int. Agrophysics*, vol. 23, no. 4, pp. 313–317, 2009.
25. L. R. Verma, R. A. Bucklin, J. B. Endan, and F. T. Wratten, "Effects of drying air parameters on rice drying models," *Trans. ASAE*, vol. 28, no. 1, pp. 296–301, 1985.
26. F. M. Botelho, P. C. Corrêa, A. Goneli, M. A. Martins, F. E. A. Magalhães, and S. C. Campos, "Periods of constant and falling-rate for infrared drying of carrot slices," *Rev. Bras. Eng. Agrícola e Ambient.*, vol. 15, no. 8, pp. 845–852, 2011.
27. U. Akyol, A. Erhan Akan, and A. Durak, "Simulation and thermodynamic analysis of a hot-air textile drying process," *J. Text. Inst.*, vol. 106, no. 3, pp. 260–274, 2015.
28. N. Kumar, B. C. Sarkar, and H. K. Sharma, "Mathematical modelling of thin layer hot air drying of carrot pomace," *J. Food Sci. Technol.*, vol. 49, no. 1, pp. 33–41, 2012.
29. D. I. Onwude, N. Hashim, R. B. Janius, N. Nawi, and K. Abdan, "Evaluation of a suitable thin layer model for drying of pumpkin under forced air convection," *Int. Food Res. J.*, vol. 23, no. 3, p. 1173, 2016.
30. S. M. Henderson, "Progress in developing the thin layer drying equation," *Trans. ASAE*, vol. 17, no. 6, pp. 1167–1168, 1974.
31. S. J. Babalis, E. Papanicolaou, N. Kyriakis, and V. G. Belessiotis, "Evaluation of thin-layer drying models for describing drying kinetics of figs (*Ficus carica*)," *J. Food Eng.*, vol. 75, no. 2, pp. 205–214, 2006.
32. M. Hamdaoui, A. Baffoun, K. Ben Chaaben, and F. Hamdaoui, "Experimental study and mathematical model to follow the drying phenomenon of knitted textile fabric," *J. Eng. Fiber. Fabr.*, vol. 8, no. 3, pp. 155–308, 2013.
33. M. Rasouli, H. R. Ghasemzadeh, and H. Nalbandi, "Convective drying of garlic (*Allium sativum L*): Part I: Drying kinetics, mathematical modeling and change in color," *Aust. J. Crop Sci.*, vol. 5, no. 13, p. 1707, 2011.
34. A. S. Kassem, "Comparative studies on thin layer drying models for wheat," in 13th International Congress on Agricultural Engineering, 1998, vol. 6, pp. 2–6.
35. K. Sacilik, R. Keskin, and A. K. Elicin, "Mathematical modelling of solar tunnel drying of thin layer organic tomato," *J. Food Eng.*, vol. 73, no. 3, pp. 231–238, 2006.
36. V. Demir, T. Gunhan, and A. K. Yagcioglu, "Mathematical modelling of convection drying of green table olives," *Biosyst. Eng.*, vol. 98, no. 1, pp. 47–53, 2007.
37. P. L. Gan and P. E. Poh, "Investigation on the effect of shapes on the drying kinetics and sensory evaluation study of dried jackfruit," *Int. J. Sci. Eng.*, vol. 7, no. 2, pp. 193–198, 2014.
38. J. R. O'callaghan, D. J. Menzies, and P. H. Bailey, "Digital simulation of agricultural drier performance," *J. Agric. Eng. Res.*, vol. 16, no. 3, pp. 223–244, 1971.
39. M. A. Hossain, J. L. Woods, and B. K. Bala, "Single-layer drying characteristics and colour kinetics of red chilli," *Int. J. Food Sci. Technol.*, vol. 42, no. 11, pp. 1367–1375, 2007.

40. Y. I. Sharaf-Eldeen, J. L. Blaisdell, and M. Y. Hamdy, "A model for ear corn drying," *Trans. ASAE*, vol. 5, no. 4, pp. 1261–1265, 1980.
41. K. K. Dash, S. Gope, A. Sethi, and M. Doloi, "Study on thin layer drying characteristics star fruit slices," *Int. J. Agric. Food Sci. Technol.*, vol. 4, no. 7, pp. 679–686, 2013.
42. E. Meisami-Asl, S. Rafiee, A. Keyhani, and A. Tabatabaeefar, "Determination of suitable thin layer drying curve model for apple slices (variety-Golab)," *Plant Omics*, vol. 3, no. 3, p. 103, 2010.
43. W. P. Da Silva, C. M. e Silva, F. J. A. Gama, and J. P. Gomes, "Mathematical models to describe thin-layer drying and to determine drying rate of whole bananas," *J. Saudi Soc. Agric. Sci.*, vol. 13, no. 1, pp. 67–74, 2014.
44. İ. Doymaz, "Evaluation of mathematical models for prediction of thin-layer drying of banana slices," *Int. J. Food Prop.*, vol. 13, no. 3, pp. 486–497, 2010.
45. K. Kaur and A. K. Singh, "Drying kinetics and quality characteristics of beetroot slices under hot air followed by microwave finish drying," *African J. Agric. Res.*, vol. 9, no. 12, pp. 1036–1044, 2014.
46. İ. Doymaz, "Evaluation of some thin-layer drying models of persimmon slices (Diospyros kaki L.)," *Energy Convers. Manag.*, vol. 56, pp. 199–205, 2012.
47. M. S. Zenoozian, H. Feng, S. M. A. Razavi, F. Shahidi, and H. R. Pourreza, "Image analysis and dynamic modeling of thin-layer drying of osmotically dehydrated pumpkin," *J. Food Process. Preserv.*, vol. 32, no. 1, pp. 88–102, 2008.
48. I. Doymaz, "The kinetics of forced convective air-drying of pumpkin slices," *J. Food Eng.*, vol. 79, no. 1, pp. 243–248, 2007.
49. K. Sacilik, "Effect of drying methods on thin-layer drying characteristics of hull-less seed pumpkin (*Cucurbita pepo* L.)," *J. Food Eng.*, vol. 79, no. 1, pp. 23–30, 2007.
50. M. S. Chhinnan, "Evaluation of selected mathematical models for describing thin-layer drying of in-shell pecans," *Trans. ASAE*, vol. 27, no. 2, pp. 610–615, 1984.
51. C. Kumar, M. U. H. Joardder, T. W. Farrell, and M. A. Karim, "Investigation of intermittent microwave convective drying (IMCD) of food materials by a coupled 3D electromagnetics and multiphase model," *Dry. Technol.*, vol. 36, no. 6, pp. 736–750, 2018
52. G. D. Mercali, I. C. Tessaro, C. P. Z. Noreña, and L. D. F. Marczak, "Mass transfer kinetics during osmotic dehydration of bananas (*Musa sapientum*, shum.)," *Int. J. Food Sci. Technol.*, vol. 45, no. 11, pp. 2281–2289, 2010.
53. I. L. Pardeshi, S. Arora, and P. A. Borker, "Thin-layer drying of green peas and selection of a suitable thin-layer drying model," *Dry. Technol.*, vol. 27, no. 2, pp. 288–295, 2009.
54. W. P. da Silva, C. M. e Silva, J. A. R. de Sousa, and V. S. O. Farias, "Empirical and diffusion models to describe water transport into chickpea (*Cicer arietinum* L.)," *Int. J. Food Sci. Technol.*, vol. 48, no. 2, pp. 267–273, 2013.
55. G. E. Page, "Factors influencing the maximum rates of air drying shelled corn in thin layers," Purdue Univ., West Lafayette, Indiana, 1949.
56. S. M. Hendorson, "Grain drying theory (I) temperature effect on drying coefficient," *J. Agric. Eng. Res.*, vol. 6, no. 3, pp. 169–174, 1961.
57. Z. Erbay and F. Icier, "A review of thin layer drying of foods: Theory, modeling, and experimental results," *Crit. Rev. Food Sci. Nutr.*, vol. 50, no. 5, pp. 441–464, 2010.
58. E. K. Akpinar, "Mathematical modelling of thin layer drying process under open sun of some aromatic plants," *J. Food Eng.*, vol. 77, no. 4, pp. 864–870, 2006.
59. İ. Doymaz, "Drying characteristics and kinetics of okra," *J. Food Eng.*, vol. 69, no. 3, pp. 275–279, 2005.
60. H. O. Menges and C. Ertekin, "Mathematical modeling of thin layer drying of Golden apples," *J. Food Eng.*, vol. 77, no. 1, pp. 119–125, 2006.

6 Single-Phase Diffusion Model

6.1 INTRODUCTION

In the previous chapter, we discussed the empirical model of drying kinetics. From this chapter and onward, different kinds of physics-based models will be discussed in detail. The single-phase model considers the sample as single-phase materials, especially the migrating water treated as a single-phase entity. Moreover, a single dominating internal moisture transfer phenomenon, namely, diffusion, is taken into consideration in the single-phase model, as shown in Figure 6.1. Due to this, the single-phase model is often referred to as a diffusion model.

The diffusion models are very popular because of their simplicity and good predictive capability [1,2]. Moreover, this type of model assumes conductive heat transfer for energy conservation. These models need the effective value of physicochemical properties including diffusivity, conductivity, specific heat and density. These effective properties are determined experimentally. In brief, an effective diffusivity-based single-phase convective drying model can be developed where the energy balance is characterized by Fourier's law of conductivity, and the moisture flux is considered due to Fickian diffusion.

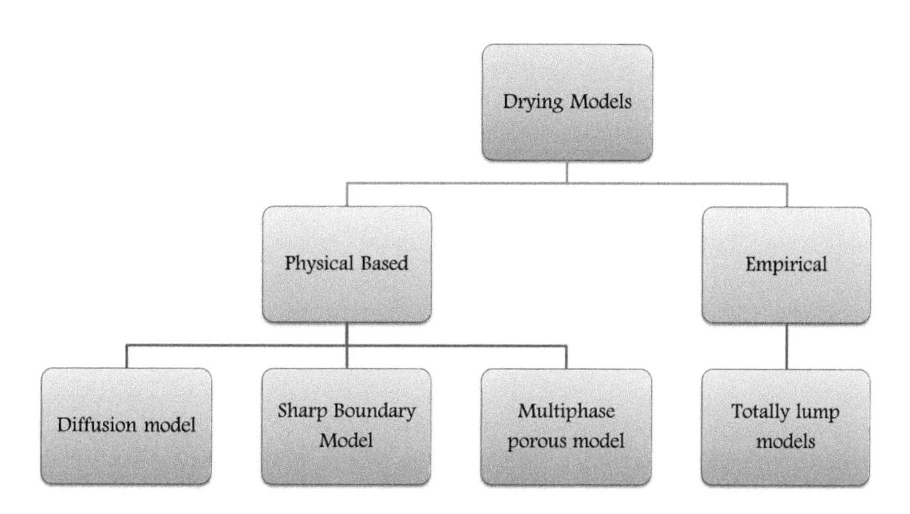

FIGURE 6.1 Classification of macro-scale drying model.

DOI: 10.1201/9780429461040-6

6.2 MODEL DEVELOPMENT

6.2.1 Geometry and Meshing

The geometry of the model needs to be developed in an optimized way. Both accuracies in sample geometrical description and computational cost need to be taken into consideration. In general, higher dimension requires more computational time and energy and makes the model the model more complex. For example, 2D geometry requires less computational cost and energy than a 3D object. Therefore, making a 3D geometry into a 2D model domain is preferable when it is possible, as shown in Figure 6.2. For example, taking the axisymmetric domain reduces geometrical complexity and the number of nodes, in turn decreasing computational energy consumption and time.

In this instance, the model developed in single-phase convection drying considers the cylindrical geometry of the sample. Other geometrical shapes can also be used instead of this according to the sample dimension.

6.2.1.1 Meshing Grid Dependency

Meshing includes the shape, size, and number of elements of a model domain. Spatial discretization needs to define these parameters. Before computing the numerical model in finite element method (FEM), finite difference method (FDM), or finite volume method (FVM), "mesh convergence" and "grid-independent"

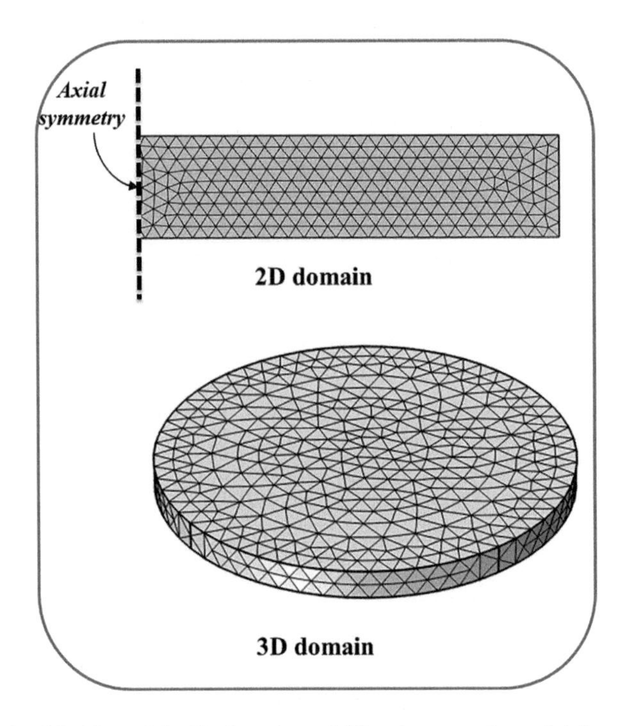

FIGURE 6.2 Meshing of the 3D domain and 2D axisymmetric model domain.

study are essential steps. Generally, the higher number of elements of a specific model domain results in higher accuracy of the simulation while taking significantly higher computational time and cost. A grid-independent test offers to determine the optimum number of elements without sacrificing the accuracy of the computational solution.

6.2.2 Assumptions

Natural phenomena are not so simple that a single model can capture all of the associated physics. There are many constrains in considering all of the fine details relating to a physical phenomenon in the simulation of mathematical modelling. Simplification is the key to run a computation model successfully without compromising the accuracy. Lack of availability of related material properties, complexity in mathematical model formulation, and limitation in computation facilities are among the main reasons for consideration of assumptions. Most often, a set of assumptions are taken into account in diffusion-based drying models as outlined below:

- The sample is initially thermo-physically homogeneous. For example, initial temperature and moisture content are the same within the entire sample. Moreover, the sample maintains constant thermo-physical properties, such as density and thermal conductivity.
- The thermo-physical properties vary with the moisture content of the material and temperature during drying phenomena.
- Convection heat transfer from the surface towards the surrounding of the sample takes place predominantly. Values of heat and mass transfer coefficients (h_m, h_T) are assumed as constant over the entire surface of the sample, and an average value based on a flat plate empirical correlation is used.
- Moisture from the inside of the sample migrates by the diffusion mechanism towards the surface.
- The heat transfer inside the sample takes place by conduction mode only.
- No heat generation is considered unless volumetric heat generation such as a microwave heating source is incorporated in the drying system.
- Radiation heat transfer at the surface is often neglected in convection drying.
- Deformation in the form of both shrinkage and expansion is disregarded in most of the conventional single-phase drying models.

6.2.3 Governing Equations

Drying is simultaneously a heat and mass transfer phenomenon. In the governing equations in diffusion-based model, only diffusion of both heat and mass transfer is considered. Therefore, convection terms are omitted from the energy and mass conservation energy.

6.2.3.1 Heat Transfer

The heat energy balance is characterized by the Fourier flux equation with null heat generation and convection terms with the following expression:

$$\rho c_p \frac{\partial T}{\partial t} = k\left(\frac{1}{r}\frac{\partial}{\partial r}\left(r\frac{\partial T}{\partial r}\right) + \frac{1}{r^2}\frac{\partial^2 T}{\partial \theta^2} + \frac{\partial^2 T}{\partial z^2}\right) \tag{6.1}$$

Here, T is the temperature (°K), ρ is the density of sample (kg/m³), c_p is the specific heat (J/kg/K), and k is the thermal conductivity (W/m/K). The equation can also be presented in Cartesian coordinate for this same modelling domain.

6.2.3.2 Mass Transfer

The mass flux of moisture is due to Fickian diffusion; therefore, this can be presented as follows:

$$\frac{\partial C}{\partial t} - \frac{1}{r}\frac{\partial}{\partial r}\left(rD_{\text{eff}}\frac{\partial C}{dr}\right) = 0 \tag{6.2}$$

where C is the moisture concentration (mol/m³) and D_{eff} is the effective diffusion coefficient (m²/s).

6.2.4 Initial and Boundary Conditions

The abovementioned transport equations deal with the element inside the medium, and they remain unchanged at any conditions on the surfaces of the medium. The governing equations do not consider the conditions of the surface; eventually, the description of heat and mass transfer phenomena remains incomplete. Therefore, the mathematical expression of boundary conditions is required, as shown in Figure 6.3, to complete overall transport phenomena.

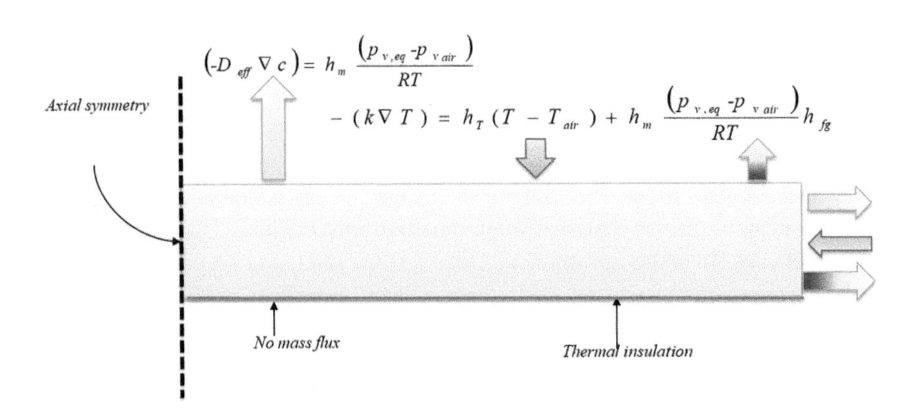

FIGURE 6.3 Heat and mass transfer boundary conditions.

6.2.4.1 Heat Transfer Boundary Conditions

Both convection and evaporation were considered at the transport boundaries, as shown in Figure 6.3. Heat transfer boundaries can be expressed by the following expression:

$$\mathbf{n}.(-k\nabla T) = h_T(T - T_{\text{air}}) - h_m \frac{(p_{v,\text{eq}} - p_{v\ \text{air}})}{RT} h_{\text{fg}} \tag{6.3}$$

where h_T is the heat transfer coefficient (W/m²/K), h_m is the mass transfer coefficient (m/s), T_{air} is the drying air temperature (°C), and h_{fg} is the latent heat of evaporation (J/kg). The heat transfer boundary condition at the symmetry boundary is given by,

$$\mathbf{n}.(-k\nabla T) = 0. \tag{6.4}$$

6.2.4.2 Mass Transfer Boundary Condition

The convection mass transfer boundary condition at the transport boundaries is given by,

$$(-D_{\text{eff}}\nabla c) = h_m \frac{(p_{v,\text{eq}} - p_{v\ \text{air}})}{RT} \tag{6.5}$$

The mass transfer boundary condition at the symmetry boundary is given by,

$$\mathbf{n}.(\mathbf{D}\nabla \mathbf{c}) = 0. \tag{6.6}$$

6.3 INPUT PARAMETERS

Input parameters consist of the information of the sample and drying conditions and can be classified as follows:

- **Sample dimension:** This is pretty much straightforward and involves the determination, in most cases, of parameters of the sample as it is sliced into a specific shape. However, most of the whole raw samples are irregular in shape.
- **Drying conditions and related properties:** Generally, the condition of the surrounding of the sample denotes the temperature, pressure, and relative humidity in conventional drying. Apart from these, related conditions need to be considered when dealing with other types of drying.
- **Thermo-physical properties of the sample:** Material properties related to heat and mass transfer need to be provided in the material models to compute the associated governing equations.

Most of the properties are not constant; rather, they vary with moisture content and temperature. To simplify, constant properties are used in many computation models that provide less accurate predictions.

The input parameters of the model are listed in Table 6.1, and some of the parameters are discussed later in this section.

Apart from the parameters mentioned in the table, there are some parameters relating to transport phenomena and those demand brief discussion in the interest of making it generic in nature.

TABLE 6.1
Input Parameters for the Model (Values Are for Typical Food Materials During Drying)

Parameter	Value	Reference
	Sample dimension	
Sample diameter, d	40 mm	Typical sample dimension
Sample thickness, b	10 mm	
	Drying conditions	
Initial temperature, T_0	303K	Typical drying condition [3]
Vapour mass fraction, W_v	0.026	
Drying air temperature, T_{air}	333K	
Universal gas constant, R_g	8.314	
Molecular weight of water, M_w	18.016 g mol^{-1}	
Latent heat of evaporation, h_{fg}	2.26e6 J kg^{-1}	
Ambient vapour pressure, $p_{v,air}$	2992 Pa	
Air, ρ_v	Ideal gas law, kg m^{-3}	
Air, C_{pa}	1005.68 J kg^{-1}K^{-1}	
Air, k_{air}	0.026 W m^{-1}K^{-1}	
Heat transfer coefficient, h_T	16.746 W/(m^{2}K)	
Mass transfer coefficient, h_m	1.79×10^{-2} m/s	
	Thermo-physical properties	
Specific heat		
Apple, C_{papple}	$C_{p\ \text{apple}} = 1000\left(1.4 + 3.22\,M_d\right)$ J kg^{-1} K^{-1}	[5, 15]
	Thermal conductivity	
Apple k_{th}	$K_{\text{apple}} = 0.49 - 0.443e^{-0.206M_d}$	
Water, $k_{th,w}$	0.644 W m^{-1}K^{-1}	
Apple solid, ρ_s	1,419 kg m^{-3}	
Initial bulk density, ρ_{apple0}	842 kg m^{-3}	
Water, ρ_w	1,000 kg m^{-3}	
Particle density	$\rho_p = \dfrac{m_s + m_w}{V_s + V_w} = \dfrac{1+X}{\dfrac{1}{\rho_s} + \dfrac{X}{\rho_w}}$	

6.3.1 Equilibrium Vapour Pressure

In mass transfer boundary conditions, setting vapour pressure-related terms is easier than setting concentration terms. Therefore, saturation vapour pressure is required in order to obtain the value of atmospheric vapour pressure at a certain relative humidity and equilibrium vapour pressure of the sample at any moisture content. Saturation vapour pressure (P_{sat}) can be obtained from the following relationships [4]:

$$P_{v,\text{sat}} = \exp\begin{bmatrix} -5800.2206/T + 1.3915 - 0.0486T + 0.4176\times10^{-4}T^2 \\ -0.01445\times10^{-7}T^3 + 6.656\ln(T) \end{bmatrix} \tag{6.7}$$

The vapour pressure of the porous materials is assumed to always be in equilibrium with the vapour pressure given by an appropriate sorption isotherm. The relationships need to be attained from experimental observation for an individual sample. For apple, the correlation of equilibrium vapour pressure with moisture and temperature is given by the following relationship [5]:

$$P_{v,\text{eq}} = P_{v,\text{sat}}(T)\exp\left(-0.182M_{db}{}^{-0.696} + 0.232e^{-43.949M}M_{db}{}^{0.0411}\ln\left[P_{\text{sat}}(T)\right]\right) \tag{6.8}$$

The relationship is for a specific sample and therefore cannot be directly applied in other porous materials. For any porous material, equilibrium vapour pressure can be obtained from sorption isotherm. Many methods, including gravimetric, manometric, and hygrometric ones, are available for determination of water sorption isotherm.

6.3.2 Effective Moisture Diffusivity

Diffusivity is commonly used to describe drying kinetics of materials in their falling rate stage, and the driving force of diffusion is the concentration gradient. This falling rate period of drying can be modelled using Fick's law.

$$\frac{\partial MR}{\partial t} = \nabla\left[D_{\text{eff}}(\nabla MR)\right] \tag{6.9}$$

where moisture ratio, MR, is calculated using the following equation:

$$MR = \frac{M - M_e}{M_0 - M_e} \tag{6.10}$$

It is assumed that the initial moisture content is uniform and that external mass transfer resistance is negligible and that moisture migration from the food material occurs in one dimension. For one-directional drying in an infinite slab, Crank [6] gave an analytical solution, as given below:

$$MR = \frac{8}{\pi^2}\sum_{n=0}^{\infty}\frac{1}{(2n+1)^2}\exp\left(-\frac{(2n+1)^2\pi^2D_{\text{eff}}t}{4L^2}\right) \tag{6.11}$$

where n is a positive integer, t is the drying time (seconds), and L is the thickness of the slab if drying occurs only on one large face. In this study, drying occurred on two faces, as slabs were placed on a mesh tray. This equation could be further simplified into Equation (6.12) by taking the first term of a series solutions as follows [7]:

$$\ln\left(\frac{M-M_e}{M_o-M_e}\right)=\ln\left(\frac{8}{\pi^2}\right)-\left(\frac{\pi^2 D_{\text{eff}}}{4L^2}t\right) \tag{6.12}$$

Effective moisture diffusion could be determined from the slope (k) obtained from the plot of $\ln\left(\frac{M-M_e}{M_o-M_e}\right)$ versus time [8]. The slope of the straight line can be expressed as follows:

$$\text{slope}=\frac{\pi^2 D}{4L^2}=K \tag{6.13}$$

Calculation of the effective diffusivity is crucial for drying models because it is the main parameter that controls the process, with a higher diffusion coefficient implying increased drying rate. The diffusion coefficient changes during drying due to the effects of sample temperature and moisture content [9]. Alternatively, some authors considered effective diffusivity as a function of shrinkage or moisture content [10], whereas others postulated it as temperature-dependent [11].

In the case of a temperature-dependent effective diffusivity value, the diffusivity increases as drying progresses. On the other hand, effective diffusivity decreases with time in the case of shrinkage or moisture dependency. This latter behaviour is ascribed to the diffusion rate decreasing as moisture gradient drops. However, Baini and Langrish [12] mentioned that shrinkage also tends to reduce the path length for diffusion, which results in increased diffusivity. Consequently, there are two opposite effects of shrinkage on effective diffusivity that theoretically may cancel each other out.

6.3.3 Temperature-Dependent Effective Diffusivity Calculation

Temperature-dependent effective diffusivity is calculated from the Arrhenius-type relationship, as shown below:

$$D_{\text{eff}}(T)=D_o\exp\left(-\frac{E}{R_g T}\right) \tag{6.14}$$

where D_o is the Arrhenius factor (m^2/s), E is the activation energy (kJ/mol), and R_g is the molar gas constant (kJ/mol/K). The effect of temperature on effective diffusivity is presented in Figure 6.4.

The activation energy was calculated from the slope of a $\ln(D_{\text{ref}})$ versus ($1/T$) graph (Figure 6.4), resulting in the values D_o=2.47×10^{-06} and E_a=36.5 kJ/mol.

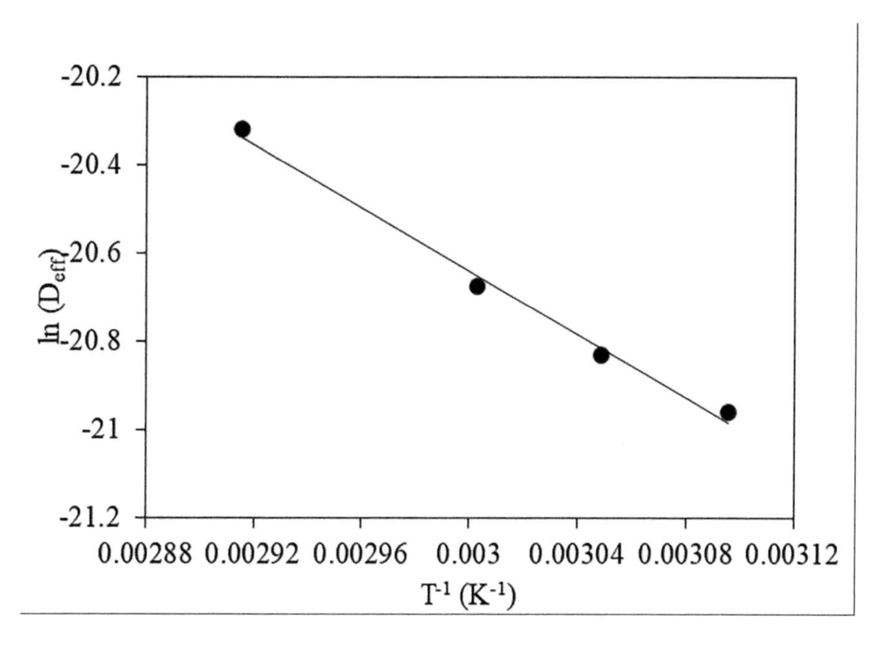

FIGURE 6.4 The effect of temperature on effective diffusivity of the apple sample.

6.3.4 Moisture-Dependent Effective Diffusivity

Moisture-dependent effective diffusivity can be calculated based on shrinkage of the sample, and the following expression can be used to calculate moisture-dependent diffusivity [13].

$$\frac{D_{\text{ref}}}{D_{\text{eff}}} = \left(\frac{b_o}{b}\right)^2 \tag{6.15}$$

where $\frac{b_o}{b}$ is called thickness ratio and D_{ref} is called the reference effective diffusivity, which is calculated by determining the slope from the experimental value, and b_0 and b are the half thickness of the material at time 0 and t, respectively.

6.3.5 Average Effective Moisture Diffusivity

Diffusivity is affected by both sample properties and process parameters. Process parameters can easily be controlled compared to sample properties. Several authors reported that the diffusion coefficient is a function of sample temperature and moisture content [9–11,14]. However, the effective diffusivity varies throughout the process due to moisture and temperature gradients. Therefore, a necessity arises to take the average of the moisture-dependent diffusivity and temperature-dependent diffusivity, as referred to in Equation (6.16), for a better prediction of effective diffusivity during drying.

$$D_{\text{eff}}(M,T) = \frac{\left(D(M) + D(T)\right)}{2} \tag{6.16}$$

After imputing all the values, we get the following moisture- and temperature-dependent effective diffusivity [15]:

$$D_{\text{eff}}(M,T) = \frac{3.81 \times 10^{-10}\left(\frac{b_o}{b}\right)^2 + 2.47 \times 10^{-06} \exp\left(-\frac{22{,}115}{R_g T}\right)}{2} \tag{6.17}$$

6.3.6 Heat and Mass Transfer Coefficient Calculation

The heat transfer coefficients are calculated from well-established correlations of Nusselt number for laminar and turbulent flow over flat plates as mentioned in the following equations:

$$Nu = \frac{h_T L}{k} = 0.664\,\text{Re}^{0.5}\,\text{Pr}^{0.33}\,(\text{Turbulent}) \tag{6.18}$$

$$Nu = \frac{h_T L}{k} = 0.0296\,\text{Re}^{0.5}\,\text{Pr}^{0.33}\,(\text{Laminar}) \tag{6.19}$$

where L is the characteristic length (m), Re is the Reynolds number, and Pr is the Prandtl number. Mass transfer coefficient is given by the Sherwood number (Sh) and the Schmidt number (Sc), respectively, as in the following relationships:

$$\text{Sh} = \frac{h_m L}{D_{va}} = 0.332\,\text{Re}^{0.5}\,\text{Sc}^{0.33}\,(\text{Turbulent}) \tag{6.20}$$

$$\text{Sh} = \frac{h_m L}{D_{va}} = 0.0296\,\text{Re}^{0.8}\,\text{Sc}^{0.33}\,(\text{Laminar}) \tag{6.21}$$

Heat and mass transfer coefficients vary with sample dimension, air velocity, and surface condition. Therefore, this coefficient needs to be determined for specific conditions. This section aim to familiarize the readers with the techniques for determining the transport coefficients. The reader interested in more insight into these topics can consult numerous other sources available, such as those listed in the references.

6.3.7 Computation

After gathering the required input parameters and initial conditions for the governing equations and boundary conditions, any computational platform (as mentioned in Chapter 4) can be used. One last thing, the time discretization, needs to be set before the solution of the model.

Similar to spatial discretization, time discretization is also a crucial step in terms of computation time and cost. Based on the applications, the time step will change. A smaller time step results in higher computation time and cost. The time step determines the speed of computation as well as the availability of the data frequency. A smaller time step takes longer time to complete the computation but provides a larger amount of temporal data. The time step is often set as a "start, step, stop" form. For example, if someone wants to compute 360 minutes of drying with a 10-minute interval, the set looks like this: (0, 10, 360). One of the vital things in determining time step is the frequency of experimental data acquisition. If the experimental data frequency is not small, then smaller time steps in computation merely cost computation time.

6.4 TYPICAL SIMULATION RESULTS

When the simulation is completed, post-processing is required to observe the drying kinetics, moisture distribution, temperature distribution, as well as variable thermo-physical properties. In this section, effective diffusivity, moisture distribution, and temperature distribution are discussed for demonstration purpose. Apart from these, many other important pieces of parameters can be investigated in detail from the simulation, provided necessary data are available.

6.4.1 Effective Moisture Diffusivity

Effective diffusivities of different kinds obtained from the simulation are compared with experimental effective diffusivity, as presented in Figure 6.5.

From the data presented in Figure 6.5, it is apparent that temperature-dependent moisture diffusivity increases with time whereas a decreasing trend is observed for moisture-dependent diffusivity. However, an average of these two effective diffusivities shows a clear agreeable trend with the effective experimental diffusivity. It is worth mentioning that when the temperature rises to reach equilibrium with the drying air, the effective moisture diffusivity starts decreasing. Consequently, a second falling drying rate is observed after this stage. Therefore, effective moisture diffusivity while considering the effect of moisture and temperature is more realistic.

6.4.2 Average Moisture Content

Figure 6.6 shows the moisture distribution throughout the sample considering diffident types of mass transfer boundary conditions. It is clear from the figure that moisture easily dries out from the outer surface of the sample. However, a relatively high concentration of moisture exists near the centre of the sample throughout the drying time [16].

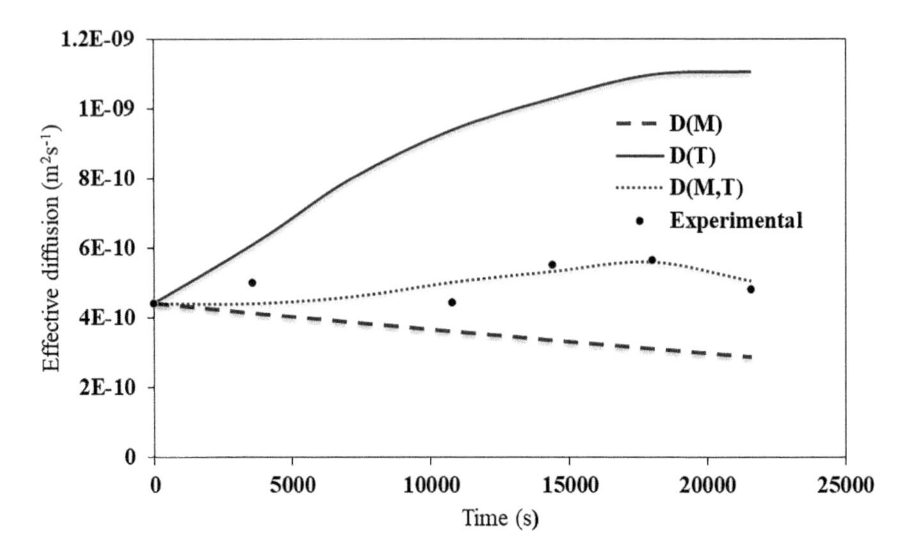

FIGURE 6.5 Change of effective diffusivity with time.

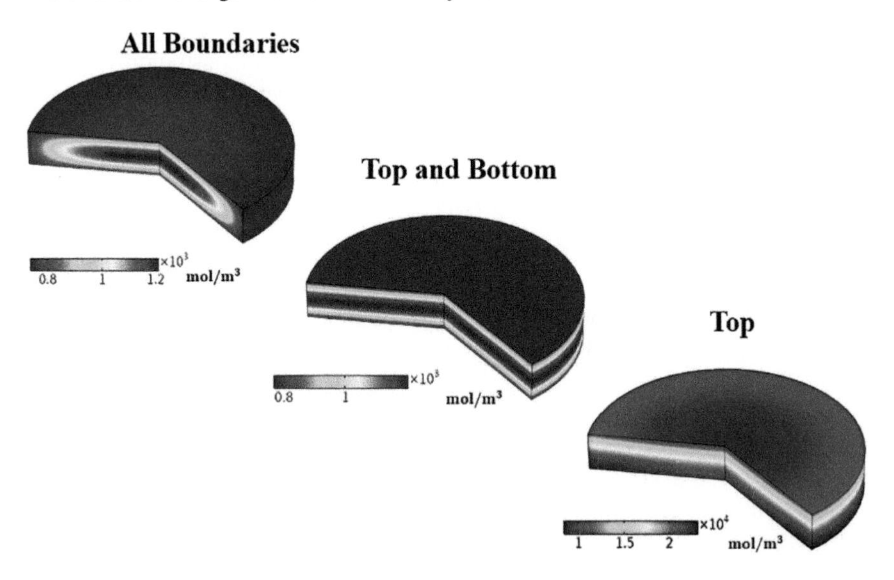

FIGURE 6.6 Water distribution within the apple sample during drying at different mass transfer boundary conditions.

To remove this high amount of water from the core, a very slow falling rate period is used to prevail in the drying phenomenon. As mentioned before, to remove the last 10% of the water from food material takes almost the same amount of time as that required for removing the first 90% of water content [17,18].

With further data analysis, spatial and temporal moisture distribution can be attained. Sometimes, an average moisture content of the sample at any time may concern someone. In those cases, further analysis of the raw data is required in some computation software, or this can even be directly obtained from some computation software.

6.4.3 Temperature Evolution

The temperature profile of the sample during drying process is shown in Figure 6.7. Temperature distribution varies with the variation of the thermal boundary condition of the sample. The surface temperature rises towards the drying temperature at a greater pace than the interior surface. When the thermal conductivity of sample, including that of biological materials, is low, significant drying time is required to make the interior temperature reach the drying temperature.

Diffusion-based single-phase model provides a wide range of information regarding the temperature and moisture distribution. Although this model can provide a good match with experimental results, it cannot provide an understanding of other transport mechanisms such as pressure-driven flow and evaporation. As there are many other internal moisture transfer phenomena taking place during drying, lumping all the water transport as diffusion cannot be justified in all situations. Therefore, multiphase models considering the transport of liquid water, vapour, and the air inside the food materials are more realistic.

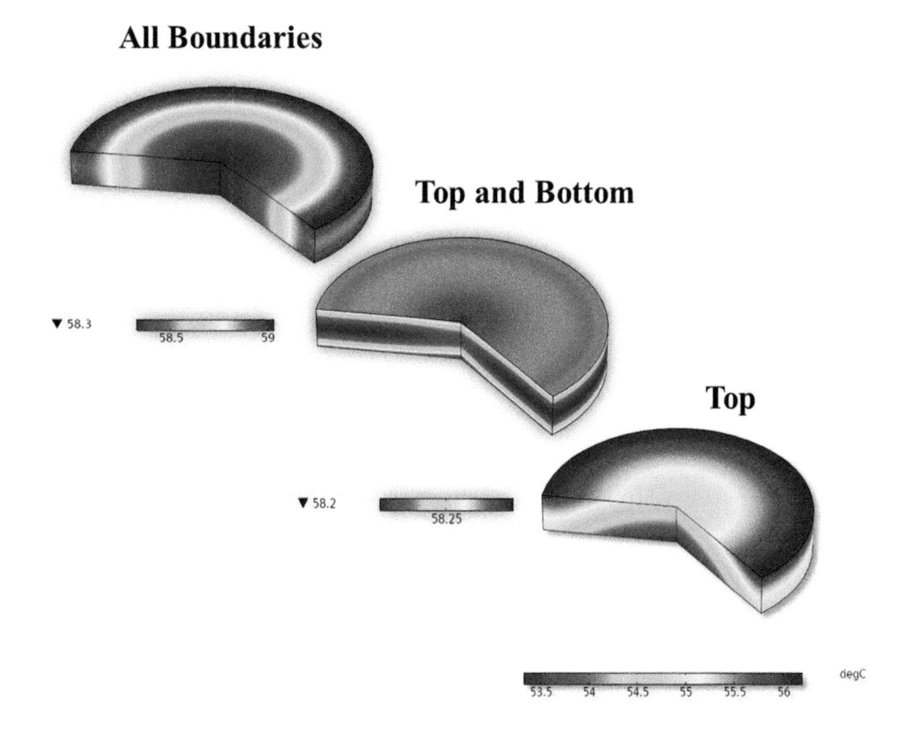

FIGURE 6.7 Temperature distribution at different boundary heating conditions.

REFERENCES

1. M. M. Rahman, M. U. Joardder, M. I. H. Khan, N. D. Pham, and M. A. Karim, "Multi-scale model of food drying: Current status and challenges," *Crit. Rev. Food Sci. Nutr.*, vol. 58, no. 5, pp. 858–876, 2018
2. C. A. Perussello, C. Kumar, F. de Castilhos, and M. A. Karim, "Heat and mass transfer modeling of the osmo-convective drying of yacon roots (Smallanthus sonchifolius)," *Appl. Therm. Eng.*, vol. 63, no. 1, pp. 23–32, 2014.
3. Y. A. Cangel and M. A. Boles, *Thermodynamics: An Engineering Approach 4th Edition in SI Units*. 2002.
4. H. Vega-Mercado, M. M. Góngora-Nieto, and G. V/ Barbosa-Cánovas, "Advances in dehydration of foods," *J. Food Eng.*, vol. 49, no. 4, pp. 271–289, 2001.
5. C. Kumar, M. U. H. Joardder, T. W. Farrell, G. J. Millar, and M. A. Karim, "Mathematical model for intermittent microwave convective drying of food materials," *Dry. Technol.*, vol. 34, no. 8, pp. 962–973, 2016.
6. C. Kumar, M. U. H. Joardder, T. W. Farrell, and M. A. Karim, "Multiphase porous media model for intermittent microwave convective drying (IMCD) of food," *Int. J. Therm. Sci.*, vol. 104, pp. 304–314, 2016.
7. A. Vega-Gálvez, M. Miranda, L. P. Díaz, L. Lopez, K. Rodriguez, and K. Di Scala, "Effective moisture diffusivity determination and mathematical modelling of the drying curves of the olive-waste cake," *Bioresour. Technol.*, vol. 101, no. 19, pp. 7265–7270, 2010.
8. M. U. H. Joardder, R. Alsbua, W. Akram, and M. A. Karim, "Effect of sample rugged surface on energy consumption and quality of plant-based food materials in convective drying," *Dry. Technol.*, pp. 1–10, 2020. DOI:10.1080/07373937.2020.1745824
9. M. Mahiuddin, M. I. H. Khan, N. Duc Pham, and M. A. Karim, "Development of fractional viscoelastic model for characterizing viscoelastic properties of food material during drying," *Food Biosci.*, vol. 23, pp. 45–53, 2018
10. M. A. Karim and M. N. A. Hawlader, "Mathematical modelling and experimental investigation of tropical fruits drying," *Int. J. Heat Mass Transf.*, vol. 48, no. 23–24, pp. 4914–4925, 2005.
11. N. Duc Pham, M. I. H. Khan, M. U. H. Joardder, A. M. N. Abesinghe, and M. A. Karim, "Quality of plant-based food materials and its prediction during intermittent drying," *Crit. Rev. Food Sci. Nutr.*, vol. 59, no. 8, pp. 1197–1211, 2019
12. R. Baini and T. A. G. Langrish, "Choosing an appropriate drying model for intermittent and continuous drying of bananas," *J. Food Eng.*, vol. 79, no. 1, pp. 330–343, 2007.
13. M. A. Karim and M. N. A. Hawlader, "Drying characteristics of banana: Theoretical modelling and experimental validation," *J. Food Eng.*, vol. 70, no. 1, pp. 35–45, 2005.
14. M. M. Rahman, M. U. H. Joardder, and A. Karim, "Non-destructive investigation of cellular level moisture distribution and morphological changes during drying of a plant-based food material," *Biosyst. Eng.*, vol. 169, pp. 126–138, 2018
15. M. U. H. Joardder, C. Kumar, and M. A. Karim, "Multiphase transfer model for intermittent microwave-convective drying of food: Considering shrinkage and pore evolution," *Int. J. Multiph. Flow*, vol. 95, pp. 101–119, 2017.
16. M. U. H. Joardder, M. Mourshed, and M. H. Masud, State of Bound Water: Measurement and Significance in Food Processing. Springer, 2019.

17. M. U. H. Joardder, A. Karim, and C. Kumar, "Better understanding of food material on the basis of water distribution using thermogravimetric analysis," in *Proceedings of the International Conference on Mechanical, Industrial and Materials Engineering (ICMIME2013)*, 2013, pp. 787–792.
18. M. U. H. Joardder, A. Karim, R. J. Brown, and C. Kumar, "Determination of effective moisture diffusivity of banana using thermogravimetric analysis," *Procedia Eng.*, vol. 90, pp. 538–543, 2014.

7 Multiphase Porous Materials Modeling

7.1 INTRODUCTION

Drying is a simultaneous heat and mass transfer problem involving multiphase species. Unlike the single-phase model, the multiphase model considers the migration of water with its liquid and gaseous phase. Therefore, a distinction exists between the transport of vapour and liquid water within the sample. Multiphase modelling allows the researchers to treat the sample as porous and hygroscopic in nature. In order to differentiate the multiphase model domain form single-phase domain, the following concepts are often referred to.

7.1.1 Porosity

Most plant-based materials, including fruits and vegetables, and many other materials can be treated as porous materials comprised of micro- and nano-scale pores trapped in a solid matrix. In other words, a material is defined as porous material when it possesses void space within a solid matrix. The void space may be filled with liquid and gaseous substances. However, the distinction of porous material from its non-porous counterpart is not easy in many cases. Porosity significantly affects the value of permeability and, therefore, the overall mass flow during drying. Significant correlation can be found between permeability and porosity. In light of the findings in the literature, materials having porosity lower than 0.25 can be classified as non-porous, intermediate porous material are those having porosity between 0.25 and 0.4, and a highly porous structure would have a porosity above 0.4 [1].

7.1.2 Tortuosity

Most of the pores in the porous sample are non-uniform in size and shape. Therefore, in order to interpret the complex nature of pore geometry, tortuosity needs to be considered [2]. Tortuosity is used to characterize the structure of porous media and generally reflects the travel path length of fluid in porous media. Water molecule often travels a path that is several times longer than the thickness of the sample. Tortuosity greatly affects the permeability and diffusivity of liquid in porous media [3].

DOI: 10.1201/9780429461040-7

7.1.3 Hygroscopic

The material can be said to be a hygroscopic material when it possesses a large amount of water physically bound with the solid matrix. For example, foods, in general, can be considered hygroscopic although there are some exceptions. These materials follow the equilibrium moisture isotherms relationships. Vapour pressure in hygroscopic material differs from the vapour pressure for pure water [4]. This pressure depends on the water activity of hygroscopic materials.

7.2 FEATURE OF MULTIPHASE DRYING MODEL

Multiphase model is significantly different in almost all of the modelling steps in comparison to single-phase diffusion-based model. From the geometry to the governing physics, there are distinct features that prevail in multiphase modelling. In the following section, we will discuss the additional concepts that are important in developing multiphase modelling [5,6]. It is worthy to note that the common concepts that have already been mentioned in the single-phase model are not repeated in this chapter.

7.2.1 Meaning of Multiphase

It is assumed that three phases, solid, liquid, and gases, are present collectively in a multiphase porous sample, as shown in Figure 7.1. The solid phase (subscript s) accounts for the solid matrix-like fibres of a cloth. The liquid phase (subscript l), mainly the liquid form of water, is counted in case of drying. Finally, the gaseous phase (subscript g) generally comprises air and water vapour. As the gaseous phase comprises of both air and water vapour, the mixture is assumed to be a perfect mixture of two ideal gases.

All of the phases are continuous and are in local thermal equilibrium. Essentially, the temperatures in all three phases are equal.

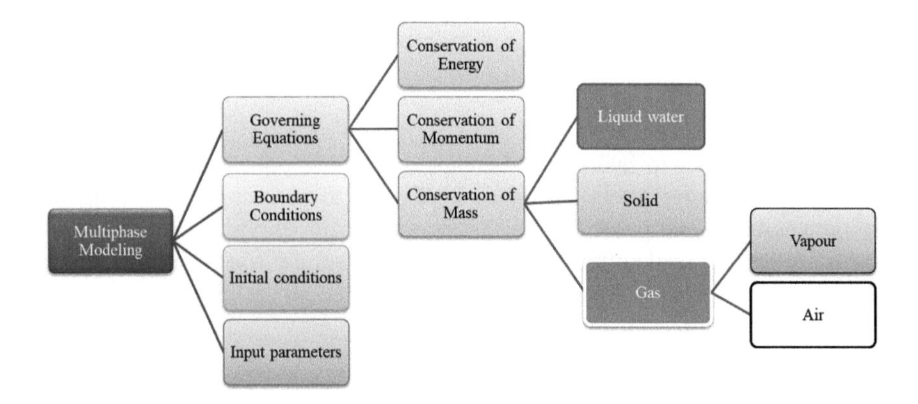

FIGURE 7.1 Multiphase modelling approach.

7.2.2 Representative Elementary Volume

Instead of treating the sample as a lump domain, it is treated as a representative elementary volume possessing effective thermophysical properties of the sample. This is the smallest volume over which an estimation can be made that will represent a quality parable of the actual sample, as shown in Figure 7.2.

7.2.3 Driving Forces of Mass Transfer

Several internal mass transfer phenomena are involved during drying, as discussed in Chapter 2. However, only diffusion mass transfer is considered in single-phase modelling, which is one of the main shortcomings of models of these kind. In multiphase modelling, liquid water transport takes place due to convective flow, capillary flow, and evaporation. However, gaseous transport takes place due to gas pressure gradients and binary diffusion. This mass transfer mechanism shows some unique feature as follows [7]:

- **Diffusion:** Molecules of any species move from higher concentration to the lower concentration. In multiphase modelling, diffusion is considered as binary diffusion and capillary diffusion.
- **Binary diffusion:** This type of diffusion deals with diffusion in a mixture of gases. Therefore, in multiphase modelling, diffusion of gases deals with binary diffusion of air and water vapour. Binary diffusion coefficient depends on the molecular weight and molecular volume of species.
- **Capillary flow:** Liquid phase often encounters capillary flow due to the capillary action of the liquid. This type of flow is also referred to as capillary diffusion. This flow is distinct from the bulk flow, which is generally caused due to gravity.
- **Pressure-driven flow:** In the case of drying where a significant pressure gradient is developed, such as microwave drying, species can flow due to this gradient of pressure. Both liquid and gas show pressure-driven flow in a porous media and often can be expressed by Darcy's law.

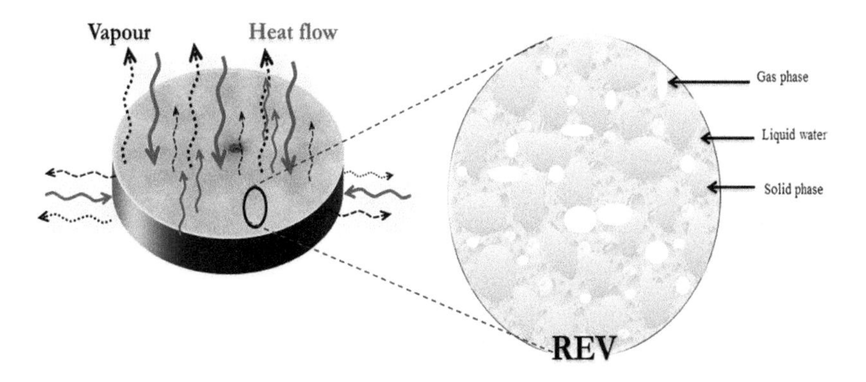

FIGURE 7.2 Representative elementary volume of multiphase sample.

- **Evaporation:** It is a phase change in heat and mass transfer process that plays an important role in drying as water migrates from the surface of the porous material to the surrounding using evaporation phenomena. Rate of evaporation often relies on the actual and equilibrium density of the vapour.

7.2.4 Assumptions

In addition to the assumptions of single-phase modelling, some of the multiphase-related assumptions need to be considered.

- All three phases coexist in thermal equilibrium. This implies that the temperature of the solid matrix is equal to the temperature of the liquid and gas phases.
- Deformation of food material in terms of shrinkage or expansion is neglected.
- The gas-phase of the porous material behaves as an ideal gas.
- The conditions of drying air, including temperature and humidity, are assumed to be in a steady state.
- The porous medium is taken as non-saturated (liquid and gas phases) because the solid matrix is usually assumed to be partially filled with water, and the rest of the volume is occupied by air.
- Effect of gravity is disregarded as the capillary force predominantly drives the liquid water within the porous sample and outweighs the effects of gravity..
- Chemical reactions that are supposed to take place are neglected.
- The non-equilibrium formulation for evaporation has been used to describe the evaporation.

7.3 GOVERNING EQUATIONS

In multiphase modelling, individual equations are required to calculate the conversion of solid, liquid, and gas species. As the amount of solid remains unchanged, three equations are required, namely, conservation of liquid, conservation water vapour, and mass fraction of air. Apart from this, in order to calculate gas pressure, Darcy's equations need to be applied. The energy and mass conservation equation are discussed in the following section [5,6]. Details of the equations related to porous media can be obtained in the pioneering work of Whitaker [8] and Bear [9].

7.3.1 Conservation of Mass

The instantaneous representative elementary volume ΔV (m^3) denotes the summation of the volume of gas, water, and solid and can be expressed as follows:

$$\Delta V = \Delta V_g + \Delta V_l + \Delta V_s \tag{7.1}$$

where ΔV_l is the volume fraction of water (m^3), ΔV_g is the volume fraction of gas (m^3), and ΔV_s is the volume fraction of solid (m^3). The apparent porosity ϕ is defined as the volume fraction occupied by water and gas. Thus,

$$\varphi = \frac{\Delta V_g + \Delta V_l}{\Delta V} \tag{7.2}$$

Liquid water saturation (S_l) and gas saturation (S_g) are defined as the fraction of pore volume occupied by the individual phase,

$$S_l = \frac{\Delta V_l}{\Delta V_l + V_g} = \frac{\Delta V_l}{\varphi \Delta V}, \tag{7.3}$$

$$\text{and } S_g = \frac{\Delta V_g}{\Delta V_l + \Delta V_g} = \frac{\Delta V_g}{\varphi \Delta V} = 1 - S_l, \tag{4}$$

The mass concentrations of liquid water (c_w), vapour (c_v), and air (c_a) in kg/m^3 are calculated by,

$$c_w = \rho_w \varphi S_l, \tag{7.5}$$

$$c_v = \frac{p_v}{RT} \varphi S_g, \tag{7.6}$$

$$\text{and } c_a = \frac{p_a}{RT} \varphi S_g, \tag{7.7}$$

where R is the universal gas constant (J/mol/K), ρ_w is the density of water (kg/m^3), p_v is the partial pressure of vapour (Pa), and p_a is the partial pressure of air (Pa).

7.3.1.1 Mass Conservation of Liquid Water

The mass conservation equation for liquid water is expressed by,

$$\frac{\partial}{\partial t}(\varphi S_l \rho_w) + \nabla \cdot (\vec{n}_w) = -\overset{*}{I} \tag{7.8}$$

where $\vec{n}_w$ is the liquid water flux (kg/m^{2}s) and $\overset{*}{I} = R_{evap}$ is the evaporation rate of liquid water to water vapour (kg/m^{3}s). The total flux of the liquid water due to the gradient of liquid pressure, $p_w = P - p_c$, is given by Darcy's law,

$$\vec{n}_w = -\rho_w \frac{k_w k_{r,w}}{\mu_w} \nabla p_w = -\rho_w \frac{k_w k_{r,w}}{\mu_w} \nabla P + \rho_w \frac{k_w k_{r,w}}{\mu_w} \nabla p_c \tag{7.9}$$

Here, P is the total gas pressure (Pa), p_c is the capillary pressure (Pa), k_w is the intrinsic permeability of water (m^2), $k_{r,w}$ is the relative permeability of water, and μ_w is the viscosity of water (Pa.s). The above expression can be from Darcy's generalized equation:

$$q = -K\frac{\partial h}{\partial z} = -K\frac{\partial}{\partial z}\left(\frac{p_w}{\rho g} + z\right) \tag{7.10}$$

If density and/or the viscosity are not constant:

$$q = -\frac{k}{\mu}\frac{\partial}{\partial z}\left(p_w + \rho g z\right) \tag{7.11}$$

The capillary pressure depends upon concentration (c_w) and temperature (T) for a particular material [10]. Therefore, further breakdown of Equation (7.9) leads to

$$\vec{n}_w = -\rho_w\frac{k_w k_{r,w}}{\mu_w}\nabla P + \rho_w\frac{k_w k_{r,w}}{\mu_w}\frac{\partial p_c}{\partial c_w}\nabla c_w + \rho_w\frac{k_w k_{r,w}}{\mu_w}\frac{\partial p_c}{\partial T}\nabla T \tag{7.12}$$

The capillary diffusivity due to the temperature gradient is known as Soret effect and is insignificant compared with the capillary diffusivity due to concentration gradients [10]. Therefore, this will be neglected in this work.

Substituting the above into Equation (7.8), we have,

$$\frac{\partial}{\partial t}\left(\varphi S_w \rho_w\right) + \nabla\cdot\left(-\rho_w\frac{k_w k_{r,w}}{\mu_w}\nabla P - \rho_w\frac{k_w k_{r,w}}{\mu_w}\frac{\partial p_c}{\partial T}\right) = -R_{evap} \tag{7.13}$$

7.3.1.2 Mass Conservation of Water Vapour

The conservation of water vapour can be written as follows:

$$\frac{\partial}{\partial t}\left(\varphi S_g \rho_g \omega_v\right) + \nabla\cdot\left(\vec{n}_v\right) = R_{evap}, \tag{7.14}$$

where ω_v is the mass fraction of vapour, ρ_g is the density of the gas (kg/m^3), and $\vec{n}_v$ is the vapour mass flux (kg/m^{2}s). As the gas is a binary mixture, $\vec{n}_v$ can be written as [11],

$$\vec{n}_v = -\rho_v\omega_v\frac{k_g k_{r,g}}{\mu_v}\nabla P - \varphi S_g \rho_g D_{eff,g}\nabla\omega_v, \tag{7.15}$$

where $k_{r,g}$ is the relative permeability of gas (m^2), k_g is the intrinsic permeability of gas (m^2), $D_{eff,g}$ is the binary diffusivity of vapour and air (m^2/s), and μ_g is the viscosity of gas (Pa.s).

7.3.1.2.1 Evaporation Rate

From the mass balance relationships and Darcy's equation, vapour pressure, water saturation, and gas pressure can be determined. However, R_{evap} remains unsolved. To solve this problem, both equilibrium and non-equilibrium descriptions of evaporation can be applied.

In case of equilibrium evaporation, the term R_{evap} cancels each other upon balancing the liquid and vapour mass conservation equation. In this multiphase porous media model, combination of the liquid water and vapour equations eliminates the evaporation rate terms. The vapour pressure p_v is assumed to be equal to the equilibrium vapour pressure, $p_{v,eq}$ and vice versa. However, equilibrium condition may be valid at the surface with lower moisture content because equilibrium condition may not be achieved due to lower moisture content at the surface during drying.

In many drying conditions, especially in fast drying, the equilibrium condition of vapour is not maintained. In such a non-equilibrium condition, it is assumed that the time scale of the evaporation process is significantly smaller than the time to reach the equilibrium. A non-equilibrium formulation as described in [12] for porous food materials can be considered to calculate the evaporation rate using the following equation:

$$R_{evap} = K_{evap}\frac{M_v}{RT}\left(p_{v,eq} - p_v\right) \tag{7.16}$$

Here, $p_{v,eq}$ is the equilibrium vapour pressure (Pa), M_v is the molecular weight of vapour (kg/mol), p_v is the vapour pressure (Pa), and K_{evap} is the material- and processes-dependent evaporation constant (s^{-1}). It is worth mentioning that K denotes evaporation rate constant and is defined as the transition of a molecule from liquid water to water vapour. For different plant-based food materials, K value is used with the range of 1,000–100,000 s^{-1} [13].

7.3.1.3 Mass Fraction of Air

After calculating the mass fraction of vapour, ω_v from water vapour mass conservation equation, the mass fraction of air, ω_a can be calculated from:

$$\omega_a = 1 - \omega_v \tag{7.17}$$

7.3.2 Continuity Equation to Solve for Pressure

The gas pressure, P, can be determined via a total mass balance for the gas phase,

$$\frac{\partial}{\partial t}\left(\rho_g \varphi S_g\right) + \nabla \cdot \left(\vec{n}_g\right) = R_{evap}, \tag{7.18}$$

where, the gas flux, $\vec{n}_g$, is given by,

$$\vec{n}_g = -\rho_g \frac{k_g k_{r,g}}{\mu_i} \nabla P. \tag{7.19}$$

Here, ρ_g is the density of the gas phase, given by,

$$\rho_g = \frac{PM_g}{RT}, \tag{7.20}$$

where M_g is the molecular weight of gas (kg/mol).

7.3.3 Energy Equation

After assuming thermal equilibrium of the phases, the energy balance equation that considers conduction and convection, radiation heat transfer, and energy sources/sinks due to evaporation/condensation can be presented as:

$$\left(\rho_a c_{p,a} + \rho_v c_{p,v}\right)\frac{\partial T}{\partial t} + \left[c_{p,a}\left(\rho_a u_g - D_{va}\rho_g \nabla \frac{\rho_a}{\rho_g}\right)\right.$$

$$\left. + c_{p,v}\left(\rho_v u_g - D_{va}\rho_g \nabla \frac{\rho_v}{\rho_g}\right)\right] \cdot \nabla T = \nabla \cdot \left(k_{eff} \nabla T\right) \tag{7.21}$$

where

$$k_{eff} = \frac{\rho_a k_a + \rho_v k_v}{\rho_a + \rho_v} \tag{7.22}$$

The energy conservation equation can also be written with other coordinating systems and considering energy source or sink.

7.3.4 Initial and Boundary Conditions

7.3.4.1 Initial Conditions

The initial conditions are required in order to solve the conservation equations. The initial conditions are considered constant throughout the sample.

$$c_{w(t=0)} = \rho_w \varphi S_{w0}, \tag{7.23}$$

$$w_{v(t=0)} = 0.0262, \tag{7.24}$$

$$P_{(t=0)} = P_{amb}, \tag{7.25}$$

$$\text{and } T_{(t=0)} = 303\text{K} \tag{7.26}$$

7.3.4.2 Boundary Condition

In general, air pressure, temperature, and relative humidity are the available conditions of the outside of the domain. On the basis of these available outside conditions, the following boundary conditions can be formulated.

7.3.4.2.1 Vapour Flux

Total vapour flux, $\vec{n}_{v,total}$, from a surface (hypothetical) in only gas phase can be written as,

$$\vec{n}_{v,total} = h_{mv} \frac{(p_v - p_{vair})}{RT}, \tag{7.27}$$

where $\vec{n}_{v,total}$ is the total vapour flux at the surface (kg/m^{2}s), h_{mv} is the mass transfer coefficient (m/s), and $p_{v,air}$ is the vapour pressure of ambient air (Pa).

In a multiphase problem, the total vapour flux on the surface encompasses evaporation from liquid water and existing vapour on the surface. The boundary conditions for liquid water and vapour phase can be written as

$$\vec{n}_w = h_{mv}\varphi S_w \frac{(p_v - p_{vair})}{RT}, \tag{7.28}$$

$$\vec{n}_v = h_{mv}\varphi S_g \frac{(p_v - p_{vair})}{RT} \tag{7.29}$$

7.3.4.2.2 Boundary Pressure and Temperature

Generally, the pressure at the boundary exposed to the environment is equal to the ambient pressure, P_{amb}. Hence, the boundary condition of pressure can be expressed as

$$P = P_{amb}. \tag{7.30}$$

7.3.4.2.3 Convection and Evaporation Thermal Boundary Condition

For the energy equation, energy can transfer by convective heat transfer and heat can be lost due to evaporation at the surface, given by,

$$q_{surf} = h_T(T - T_{air}) + h_{mv}\phi S_w \frac{(p_v - p_{vair})}{RT} h_{fg}. \tag{7.31}$$

7.4 INPUT PARAMETERS

Many of the required input parameters of the multiphase model have already been listed in Chapter 6. Additional parameters that are not listed in the single-phase model are discussed in the following sections.

7.4.1 Thermo-Physical Properties

The thermophysical properties of the REV are obtained by the volume-weighted average of the different phases and can be calculated by the following equations:

$$\rho_{eff} = \varphi\left(S_g\rho_g + S_w\rho_w\right) + (1-\varphi)\rho_s, \tag{7.32}$$

$$c_{peff} = \varphi\left(S_g c_{pg} + S_w c_{pw}\right) + (1-\varphi)c_{ps}, \tag{7.33}$$

$$\text{and } k_{eff} = \varphi\left(S_g k_{th,g} + S_w k_{th,w}\right) + (1-\varphi)k_{th,s}, \tag{7.34}$$

Here ρ_s is the solid density (kg/m³). $k_{th,g}$,$k_{th,w}$, and $k_{th,s}$ are the thermal conductivity of gas, water, and solid (W/m/K), respectively. c_{pg}, c_{pw}, and c_{ps} are the specific heat capacity of gas, water, and solid (J/kg/K), respectively. Table 7.1 shows the value of effective thermal properties of some typical foods.

The equilibrium vapour pressure of apple, as presented in Equation (7.35), can be obtained from the sorption isotherm [19].

$$p_{v,eq} = P_{v,sat}(T)\exp\left(-0.182M_{db}^{-0.696} + 0.232e^{-43.949M}M_{db}^{0.0411}\ln[P_{sat}(T)]\right) \tag{7.35}$$

Here, M_{db} is the moisture content (dry basis) and $P_{v,sat}$ is the saturated vapour pressure of water (Pa) as mentioned in Chapter 6 and it can be related to S_w via

$$M_{db} = \frac{\varphi S_w \rho_w}{(1-\varphi)\rho_s}. \tag{7.36}$$

The vapour pressure, p_v, is obtained from partial pressure relations given by

$$p_v = \chi_v P, \tag{7.37}$$

where χ_v is the mole fraction of vapour and P is the total pressure (Pa).

TABLE 7.1
Effective Thermal Properties of the Porous Sample

Sample	Effective Thermal Conductivity, W/(m·K)	Effective Specific Heat, J/(kg·K)	Ref.
Banana	$0.006M_{wb} + 0.12$	$0.811M_{wb}^2 - 24.75M_{wb} + 1742$	[14]
Potato	$0.5963 - \left(\frac{0.1931}{M_{db}}\right) + \left(\frac{0.0301}{M_{db}^2}\right)$	$4184 \times (0.406 + 0.00146T + 0.203M_{db} - 0.0249M_{db}^2)$	[15]
Carrot	$0.55M_{wb} + 0.26(1 - M_{wb})$	$1{,}000 \times \left(4.19M_{wb} + 0.84(1 - M_{wb})\right)$	[16]
Papaya	$2.673 \times 10^{-3}(T - 273) + 3.246 \times 10^{-5}$	$15.56 \times (T - 273)^2 - 8.042 \times (T - 273) + 3352$	[17]
Quince	$0.148 + 0.493M_{wb}$	$1000 \times (1.26 + 2.97M_{wb})$	[18]

The mole fraction of vapour, χ_v, can be calculated from the mass fractions and molecular weight of vapour and air as

$$\chi_v = \frac{\omega_v M_a}{\omega_v M_a + \omega_a M_v}, \tag{38}$$

where M_v is the molecular mass of vapour (kg/mol) and M_a is the molecular mass of air (kg/mol).

7.4.2 Porous Structure-Related Properties

7.4.2.1 Porosity

The porosity of a material indicates the volume of the air space present in the food product. All hygroscopic food materials have some porosity (air space), and several mathematical expressions have been proposed to predict the porosity of a material as a function of the moisture content. These models can be grouped into two categories: (i) theoretical models that are built based upon the understanding of the fundamental phenomena and mechanisms that may be involved in pore formation, and (ii) empirical models that are built by fitting the model's parameters to the experimental data. Rahman [20] demonstrated that porosity is a function of the shrinkage coefficient and developed a physics-based model for porosity measurement, which is shown in the equation below:

$$\phi = \frac{\rho_P(\varphi_i - \varphi)}{\rho_w(\varphi - 1)} \tag{7.39}$$

Very recently, Joardder et al. [21] developed a model for porosity prediction considering both material properties and process parameters. The model is as follows [21]:

$$\varphi = \frac{\varphi_0\left(\frac{T - T_g}{T_i - T_{gi}}\right) + \frac{m_i}{X_i}\left(\frac{X_{fi} - X_f}{\rho_w} + \left(1 - \frac{\rho_w}{\rho_s}\right)\frac{X_{bi} - X_b}{\rho_w}\right)}{V_i\left(1 - \left(\varphi_i\left(\frac{T - T_g}{T_i - T_{gi}}\right)\right)\right) - \frac{\rho_w}{\rho_s}\left(\frac{X_{bi} - X_b}{\rho_w}\right)\frac{m_i}{X_i}} \tag{7.40}$$

where X is the moisture content dry basis, T is the temperature (K), m is the mass (kg), V is the volume (m^3), and i is initial, f is final, b is bound, w is water, and g is the glass transition. The interested reader can go through the review of Joardder et al. [22] for more details on porosity-related relationships.

7.4.2.2 Permeability

Permeability is an important factor concerning describing water transport due to the pressure gradient in unsaturated porous media. The value of the permeability determines the extent of pressure generation inside the material. The smaller the permeability, the lower the moisture transport and the higher the internal pressure, and vice versa.

Permeability is a property of porous materials that defines the resistance to flow of a fluid through the material under the influence of a pressure gradient. According to Darcy's law for laminar flow, pressure drop and flow rate show a linear relationship through a porous material [23]. The slope of this line is related to permeability as shown below:

$$k = \frac{Q\mu L}{A\Delta P} \tag{7.41}$$

Here k is the permeability, Q is the volumetric flow rate, μ is the fluid viscosity, A is the sample cross-sectional area, and ΔP is the pressure gradient.

The permeability of a material to a fluid, k, is a product of intrinsic permeability, k_i, of the material and relative permeability, $k_{i,r}$, of the fluid to that material [9], namely,

$$k = k_i k_{i,r}. \tag{7.42}$$

The intrinsic permeability, k_i, represents the permeability of a liquid or gas in the fully saturated state. The relative permeability of a phase is a dimensionless measure of the effective permeability of that phase. It is the ratio of the effective permeability of that phase to the absolute permeability. Relative permeability of gases is a function of saturation.

The intrinsic permeability depends on the pore structure of the material and, from literature, it is found as follows [24]:

$$k_w = 5.578 \times 10^{-12} \frac{\varphi^3}{(1-\varphi)^2} \quad (0.39 < \varphi < 0.77) \tag{7.43}$$

Relative permeabilities are generally expressed as functions of liquid saturation and there are numerous studies that have developed such functions [25]. For example, the relative permeabilities of water $k_{r,w}$ and gas $k_{r,g}$ for apple were obtained from the study of Feng et al. [24] as

$$k_{r,w} = S_w^{\ 3} \tag{7.44}$$

$$\text{and } k_{r,g} = 1.01e^{-10.86 S_w} \tag{7.45}$$

7.4.2.3 Capillary Diffusivity of Liquid Water

Liquid from food materials can flow from materials of higher concentration to those at lower concentration of water due to the difference by capillary action. This is referred to as "unsaturated" flow and is extremely important in food

TABLE 7.2
Capillary Diffusivity of Liquid Water

Samples	Water Capillary Diffusivity, m^2/s	Ref.
Potato	$4.49\times10^{-5}\exp\left(-\frac{2{,}172}{T}\right)$	[26]
Carrot	$\frac{1}{3600}\exp\left(-0.97-\frac{3459.8}{T}+0.059M_{db}\right)$	[16]
Bread	$2.8945\exp\left(1.26M_{wb}-2.76M_{wb}^2+4.96M_{wb}^3-\frac{6117.4}{T}\right)$	[27]

processing, for example, in the drying of food materials. During the modelling of multiphase flow for different food processing systems, the property of capillary diffusivity D_{cap} is very important. Table 7.2 lists the typical value of capillary diffusivity of liquid water of some food materials.

7.4.3 Gas-Related Properties

7.4.3.1 The Viscosity of Water and Gas

Viscosities of water [28] and gas [29] as a function of temperature are given by

$$\mu_w = \rho_w e^{\left(-19.143+\frac{1540}{T}\right)} \tag{7.46}$$

$$\text{and } \mu_g = 0.017\times10^{-3}\left(\frac{T}{273}\right)^{0.65}. \tag{7.47}$$

7.4.3.2 Effective Gas Diffusivity

The effective gas diffusivity can be calculated as a function of gas saturation and porosity according to the Bruggeman correction [30] given by

$$D_{eff,g} = D_{va}\left(S_g\varphi\right)^{4/3} \tag{7.48}$$

Here, binary diffusivity, D_{va}, can be written as

$$D_{va} = 2.3\times10^{-5}\frac{P_0}{P}\left(\frac{T}{T_0}\right)^{1.81}, \tag{7.49}$$

where $T_0 = 256\text{K}$ and $P_0 = 1$ atm. For simplicity, in this study, effective gas diffusivity was considered as $2.6\times10^{-6}\ m^2/s$ [31].

7.4.4 Drying Air Condition (Relative Humidity)

Moist air (subscript g) is modelled as a perfect mixture of two ideal gasses, namely, dry air (subscript) and water vapour (subscript). For each component, the ideal gas law can be defined as follows [32]:

$$p_a = \rho_a R_a T = \frac{m_a}{V_g} R_a T = x_a \rho_g R_a T \tag{7.50}$$

$$p_v = \rho_v R_v T = \frac{m_v}{V_g} R_v T = x_v \rho_g R_v T \tag{7.51}$$

where p_a and p_v are partial pressures and V_g is the mixture volume. The density of individual species is related to their corresponding mass fraction, as shown below.

$$\rho_a = x_a \rho_g \tag{7.52}$$

$$\rho_v = x_v \rho_g \tag{7.53}$$

To quantify the amount of moisture in the air, relative humidity (φ) is introduced, which is expressed by the following expression:

$$\varphi = \frac{p_v}{p_{v,sat}} \tag{7.54}$$

7.5 TYPICAL SIMULATION RESULTS

Upon completion of the simulations, temporal and spatial evolution of numerous parameters and variables can be observed, which is not possible from a single-phase model. In this section, profiles of moisture, temperature, and evaporation rate are presented and discussed.

7.5.1 Average Moisture Content

The evolution of average moisture content obtained from multiphase models is shown in Figure 7.3. The result obtained from models shows that throughout drying, the water saturation near the surface was lower than that in the centre region. Moreover, due to an adequate supply of drying air, all of the water at the surface migrated through convective flow and was instantly removed from the surface.

This nature of moisture distributions within the sample seems to be consistent as the core contained higher moisture content compared to the surface. Along the drying time trajectory, average moisture content decreased with a constant rate. However, the moisture content at the centre of the sample does not decrease significantly.

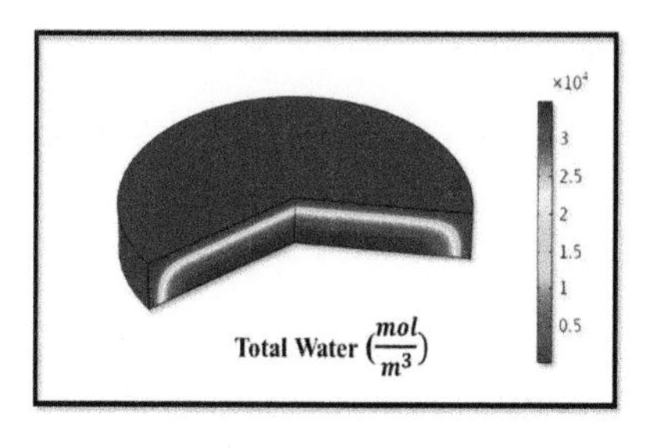

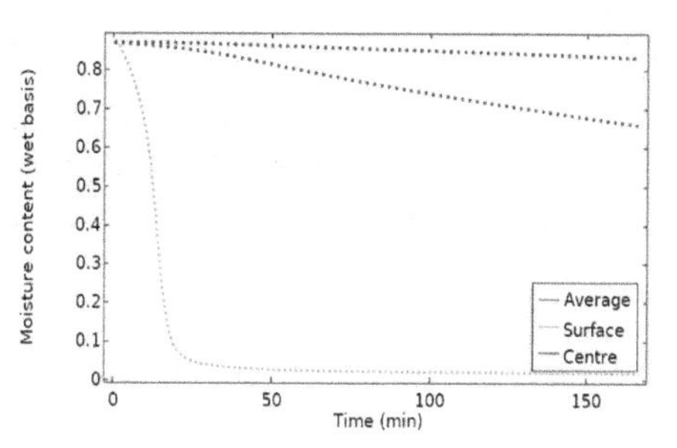

FIGURE 7.3 Moisture concentration and content during drying.

7.5.2 Liquid and Gas Saturation

As drying involves phase change phenomena, continuous change of liquid and gas saturation can be observed within the sample. Figure 7.4 shows the saturation of liquid water throughout the sample at different times. Water saturation is more obvious towards the core of the sample than its exterior. This is caused due to the conversion of liquid water into vapour upon heat transfer. Moreover, liquid saturation decreases with time for entire sample due to moisture migration towards the surroundings.

Unlike the moisture distribution, saturation was found to increase with drying time within the sample. However, the vapour saturation near the surface is higher than at the centre because the phase change-facilitated heat energy propagated slowly from the surface toward the centre. In addition, it is apparent from Figure 7.5 that the concentration of vapour at any point of the sample is decreased with time for both the models.

7.5.3 Temperature Evolution and Distribution

Figure 7.6 shows the surface temperature evolution of dried materials. Due to sudden exposure of the sample to higher air temperature, the surface temperature rose sharply at the beginning of the drying process.

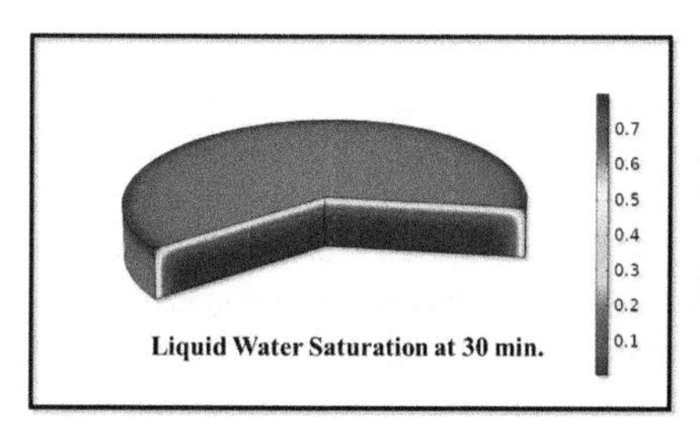

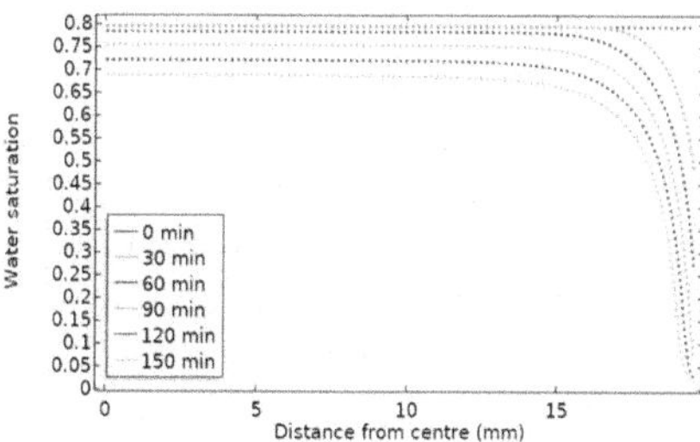

FIGURE 7.4 Temporal and spatial evolution of liquid saturation.

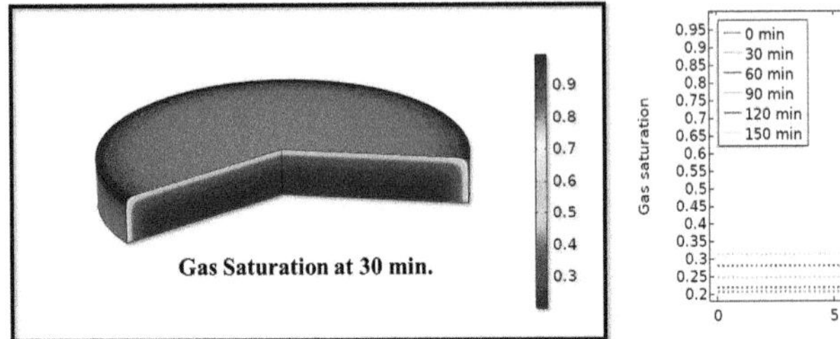

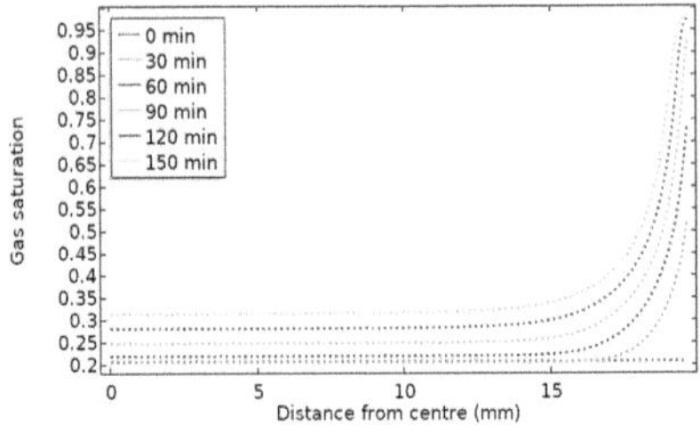

FIGURE 7.5 Spatial and temporal evolution of gas (vapour and air) saturation.

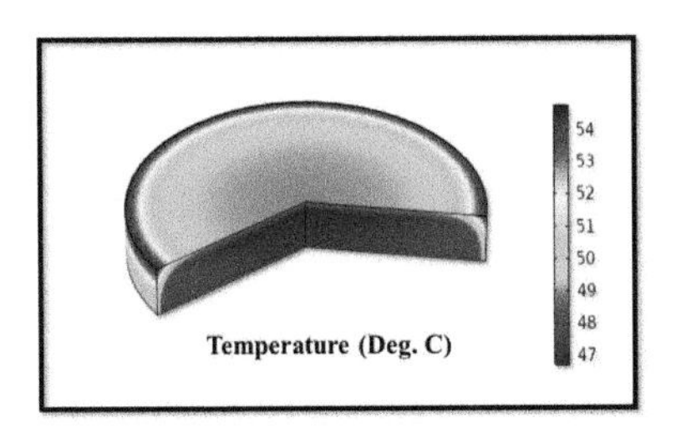

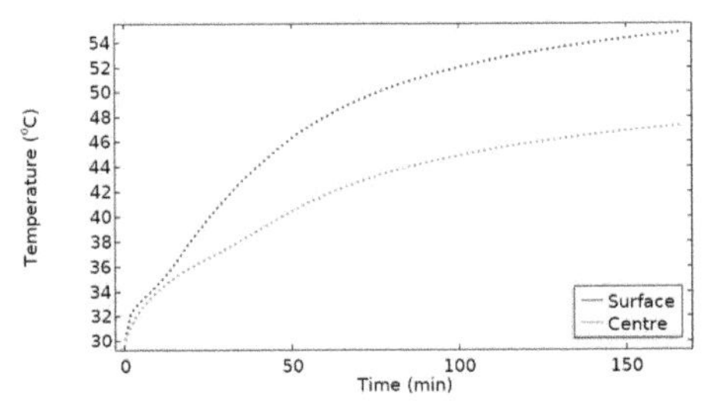

FIGURE 7.6 Temporal and spatial temperature distribution.

Moreover, the edge of the sample had the highest temperature due to consistent multidirectional heating, whereas the increment of temperature at the centre of the sample is relatively slower than that at the surface.

7.5.4 Evaporation Rate and Vapour Pressure

One of the advantages of multiphase modelling is that it can provide specific information regarding different phase of the sample. Vapour pressure and evaporation rate are two examples of such information that can be attained from multiphase modelling. Vapour pressure and evaporation rate are presented in Figure 7.7. From this figure, it is revealed that a higher evaporation rate occurs near the surface and maintains a non-equilibrium condition.

It is clear from the figure that the evaporation maintains a zone, which starts not exactly from the surface, but rather beneath the surface, and decreases to zero evaporation towards the core (as demonstrated in Figure 7.7). This result may be explained by the fact that the liquid water saturation becomes lower near the surface of the sample. This lower liquid water saturation near the surface does not contribute in the vapour pressure and equilibrium vapour pressure. Eventually, the evaporation front moves inside of the material as drying progresses.

Besides, the vapour pressure is maximum at the core region and progressively drops towards the surface. Due to lower vapour porosity, the migration of gas from the interior is restricted while the sample contains higher moisture content, and this causes a rise in total gas pressure.

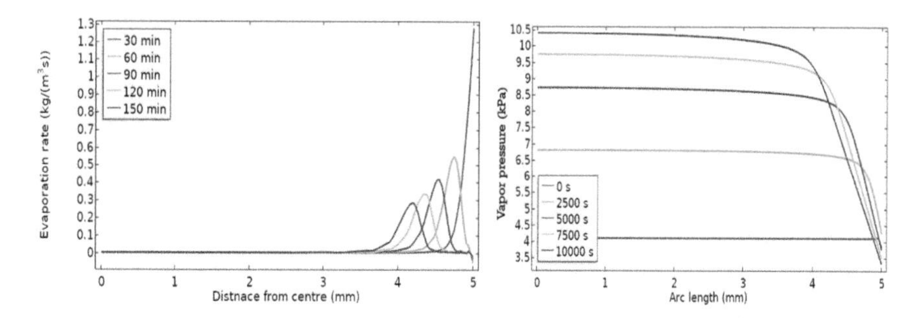

FIGURE 7.7 Spatial distribution of evaporation rate and vapour pressure at different times during drying [5,6].

7.6 CHALLENGES AND POSSIBLE SIMPLIFICATIONS

Multiphase modelling offers many advantages, including distribution of different phases within the sample. The deformable nature of porous materials and physicochemical properties are the main challenges that are encountered in the development of multiphase modelling.

7.6.1 Challenges

7.6.1.1 Shrinkage

During food processing, microstructural stress developed due to the moisture gradient within the product leads to shrinkage. When moisture migrates from a porous material, shrinkage must take place in that material [33]. During the drying time, moisture migrating from the material cell results in pore formation. Therefore, it can be said that porosity is a variable property of deformable materials. As modelling of instantaneous deformation is straightforward, most of the existing multiphase models consider porosity as a constant property. Without considering variable porosity in the multiphase model, saturations of different phases will be predicted erroneously. We will give a brief description of shrinkage modelling in Chapter 9 to familiarize the reader with the basic mechanisms involved in the deformation of porous media.

7.6.1.2 Properties of Porous Material

It is clear from the properties and parameter sections that multiphase model requires far more thermo-physical properties of porous materials. The required properties vary with sample types, and an experimental protocol is usually required to determine the properties if these are not available in the literature. Porosity, liquid water diffusivity, gas permeability, capillary pressure and, effective gas diffusivity are common but are not available for most of the porous materials. The dependency of the properties on the moisture content and temperature makes the scenario worse.

7.6.2 Simplification

At a glance, the multiphase model seems more complicated than the single-phase model. The following assumptions can be considered to make the multiphase model simpler.

- Gas pressure can be assumed constant, resulting in $p_a = (P - p_v) = \text{constant}$.
- Considering sharp moving boundaries between dry and wet regions simplifies the multiphase model to a great extent.
- Constant values of moisture-dependent variables can lead to great simplification.

Successful implementation of multiphase modelling approach can provide a more accurate prediction of moisture and temperature distribution. Observation of the evolution of different phases can be possible in multiphase modelling. Despite these advantages of multiphase modelling, the involvement of a wide range of input parameters makes it quite difficult in successful simulation. Most of the time, the new sample needs multiple experimental investigations for determining these thermo-physical properties.

REFERENCES

1. M. U. H. Joardder, A. Karim, C. Kumar, and R. J. Brown, *Porosity: Establishing the Relationship between Drying Parameters and Dried Food Quality*. Springer, 2015.
2. B. Ghanbarian, A. G. Hunt, R. P. Ewing, and M. Sahimi, "Tortuosity in porous media: A critical review," *Soil Sci. Soc. Am. J.*, vol. 77, no. 5, pp. 1461–1477, 2013.
3. L. Shen and Z. Chen, "Critical review of the impact of tortuosity on diffusion," *Chem. Eng. Sci.*, vol. 62, no. 14, pp. 3748–3755, 2007.
4. M. I. H. Khan and M. A. Karim, "Cellular water distribution, transport, and its investigation methods for plant-based food material," *Food Res. Int.*, vol. 99, pp. 1–14, 2017
5. C. Kumar, M. U. H. Joardder, T. W. Farrell, and M. A. Karim, "Investigation of intermittent microwave convective drying (IMCD) of food materials by a coupled 3D electromagnetics and multiphase model," *Dry. Technol.*, vol. 36, no. 6, pp. 736–750, 2018.
6. M. U. H. Joardder, C. Kumar, and M. A. Karim, "Multiphase transfer model for intermittent microwave-convective drying of food: Considering shrinkage and pore evolution," *Int. J. Multiph. Flow*, vol. 95, pp. 101–119, 2017.
7. M. U. H. Joardder, M. Mourshed, and M. H. Masud, *State of Bound Water: Measurement and Significance in Food Processing*. Springer, 2019.
8. S. Whitaker, "Simultaneous heat, mass, and momentum transfer in porous media: a theory of drying," in *Advances in Heat Transfer*, vol. 13, Elsevier, 1977, pp. 119–203.
9. J. Bear, *Dynamics of Fluids in Porous Media*. Courier Corporation, 2013.
10. A. K. Datta, "Porous media approaches to studying simultaneous heat and mass transfer in food processes. I: Problem formulations," *J. Food Eng.*, vol. 80, no. 1, pp. 80–95, 2007.
11. R. B. Bird, "Transport phenomena," *Appl. Mech. Rev.*, vol. 55, no. 1, pp. R1–R4, 2002.

12. H. Ni, A. K. Datta, and K. E. Torrance, "Moisture transport in intensive microwave heating of biomaterials: a multiphase porous media model," *Int. J. Heat Mass Transf.*, vol. 42, no. 8, pp. 1501–1512, 1999.
13. A. Halder, A. Dhall, and A. K. Datta, "Modelling transport in porous media with phase change: applications to food processing," *J. Heat Transfer*, vol. 133, pp. 1–13, 2011.
14. M. I. H. Khan, C. Kumar, M. U. H. Joardder, and M. A. Karim, "Determination of appropriate effective diffusivity for different food materials," *Dry. Technol.*, vol. 35, no. 3, pp. 335–346, 2017.
15. P. P. Tripathy and S. Kumar, "Modelling of heat transfer and energy analysis of potato slices and cylinders during solar drying," *Appl. Therm. Eng.*, vol. 29, no. 5–6, pp. 884–891, 2009.
16. B. A. Souraki, A. Andres, and D. Mowla, "Mathematical Modelling of microwave-assisted inert medium fluidized bed drying of cylindrical carrot samples," *Chem. Eng. Process. Process Intensif.*, vol. 48, no. 1, pp. 296–305, 2009.
17. R. A. Lemus-Mondaca, A. Vega-Gálvez, C. E. Zambra, and N. O. Moraga, "Modelling 3D conjugate heat and mass transfer for turbulent air drying of Chilean papaya in a direct contact dryer," *Heat Mass Transf.*, vol. 53, no. 1, pp. 11–24, 2017.
18. D. A. Tzempelikos, D. Mitrakos, A. P. Vouros, A. V Bardakas, A. E. Filios, and D. P. Margaris, "Numerical Modelling of heat and mass transfer during convective drying of cylindrical quince slices," *J. Food Eng.*, vol. 156, pp. 10–21, 2015.
19. C. Ratti, G. H. Crapiste, and E. Rotstein, "A new water sorption equilibrium expression for solid foods based on thermodynamic considerations," *J. Food Sci.*, vol. 54, no. 3, pp. 738–742, 1989.
20. M. S. Rahman, *Food Properties Handbook*. CRC Press, 2009.
21. M. I. H. Khan, M. U. H. Joardder, C. Kumar, and M. A. Karim, "Multiphase porous media modelling: A novel approach to predicting food processing performance," *Crit. Rev. Food Sci. Nutr.*, vol. 58, no. 4, pp. 528–546, 2018.
22. M. U. H. Joardder, C. Kumar, and M. A. Karim, "Prediction of porosity of food materials during drying: Current challenges and directions," *Crit. Rev. Food Sci. Nutr.*, vol. 58, no. 17, pp. 2896–2907, 2018.
23. A. E. Scheidegger, *The Physics of Flow Through Porous Media*. University of Toronto Press, 1957.
24. H. Feng, J. Tang, O. A. Plumb, and R. P. Cavalieri, "Intrinsic and relative permeability for flow of humid air in unsaturated apple tissues," *J. Food Eng.*, vol. 62, no. 2, pp. 185–192, 2004.
25. O. A. Plumb, "Heat transfer during unsaturated flow in porous media," in *Convective Heat and Mass Transfer in Porous Media*, Springer, 1991, pp. 435–464.
26. J. Chen, K. Pitchai, S. Birla, M. Negahban, D. Jones, and J. Subbiah, "Heat and mass transport during microwave heating of mashed potato in domestic oven—model development, validation, and sensitivity analysis," *J. Food Sci.*, vol. 79, no. 10, pp. E1991–E2004, 2014.
27. T. Gulati, A. K. Datta, C. J. Doona, R. R. Ruan, and F. E. Feeherry, "Modelling moisture migration in a multi-domain food system: Application to storage of a sandwich system," *Food Res. Int.*, vol. 76, pp. 427–438, 2015.
28. S. L. Truscott and I. W. Turner, "A heterogeneous three-dimensional computational model for wood drying," *Appl. Math. Model.*, vol. 29, no. 4, pp. 381–410, 2005.
29. T. Gulati and A. K. Datta, "Enabling computer-aided food process engineering: property estimation equations for transport phenomena-based models," *J. Food Eng.*, vol. 116, no. 2, pp. 483–504, 2013.

30. H. Ni, "Multiphase moisture transport in porous media under intensive microwave heating," Cornell University, January, 1997.
31. A. K. Datta, "Porous media approaches to studying simultaneous heat and mass transfer in food processes. II: Property data and representative results," *J. Food Eng.*, vol. 80, no. 1, pp. 96–110, 2007.
32. S. A. Pemberton, "A novel approach to multiphysics Modelling of heat and mass transfer in porous media," University of Tennessee, 2013.
33. M. U. H. Joardder, C. Kumar, and M. A. Karim, "Food structure: Its formation and relationships with other properties," *Crit. Rev. Food Sci. Nutr.*, vol. 57, no. 6, pp. 1190–1205, 2017.

8 Micro-Scale Drying Model

8.1 INTRODUCTION

Length scale and time scale play vital roles in most of the physical phenomena. Changing length scale may affect heat and mass transfer during drying [1]. In other words, drying at micro-scale may vary from its macro-scale counterpart in terms of transport properties, boundary conditions, and even in governing conditions. Eventually, moisture and temperature distribution vary significantly at the micro-level. In the previous chapter, we have described the empirical and macro-scale models, shown in Figure 8.1.

Micro-scale modelling allows the observation of simultaneous heat and mass transfer at the smaller functional level such as the cells of plant-based food materials [2]. Micro-scale model is significantly different in many aspects in comparison to a macro-scale model. From geometry to governing physics, there are distinct features that prevail in micro-scale modelling. In the following section, we will discuss the additional concepts that are important in developing microscale modelling [2,3].

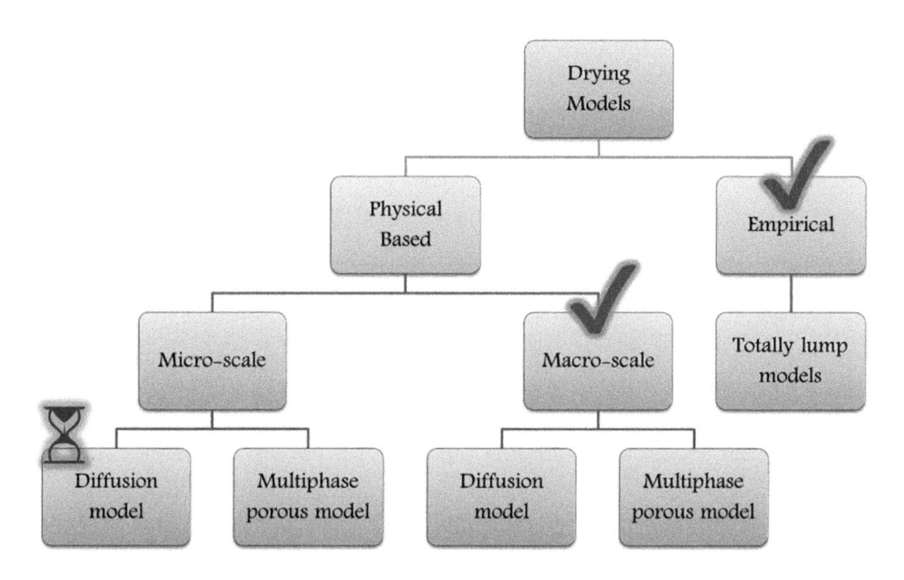

FIGURE 8.1 Micro-scale modelling approach during drying.

DOI: 10.1201/9780429461040-8

8.1.1 Defining Micro-Scale

Micro-level changes during drying are critical for developing an optimum drying process. Microstructural characteristics of the material are the determinant physical properties that profoundly affect heat and mass transfer during drying. Definition of micro-scale is not an easy task; rather, it depends on the context. In the case of drying, homogenous materials do not require the consideration of micro-scale drying phenomena while porous heterogeneous materials show different features at different length scales. There are different types of pores and capillaries (Figure 8.2) that exist in porous materials, and the determination of the size of micro-scale depends on the nature of these pore sizes.

The cell-level model of plant-based materials can be treated as a micro-scale model. However, this modelling approach is not beneficial for the materials that do not show different transport phenomena from their macro-scale counterparts. The general consideration of the micro-scale models are the heterogeneous structure and diverse transport phenomena at different length scales that facilitate micro- and macro-level modelling.

8.1.2 Micro-Scale Domain

Micro-level structural properties are important in heat and mass transfer, and thus the drying kinetics. The structural component at different length scale participates in transport phenomena with different characteristics [4]. For cellular materials, transport phenomena that take place through the pores, cells, and cell walls of food materials can be categorized as the micro-scale phenomena, as shown in Figure 8.3.

Cellular-level structural parameters such as diameter, wall thickness, shape, and organization of the cells influence the tissue-level properties of food [6]. The geometrical properties of the micro-scale, including pore diameter and pore

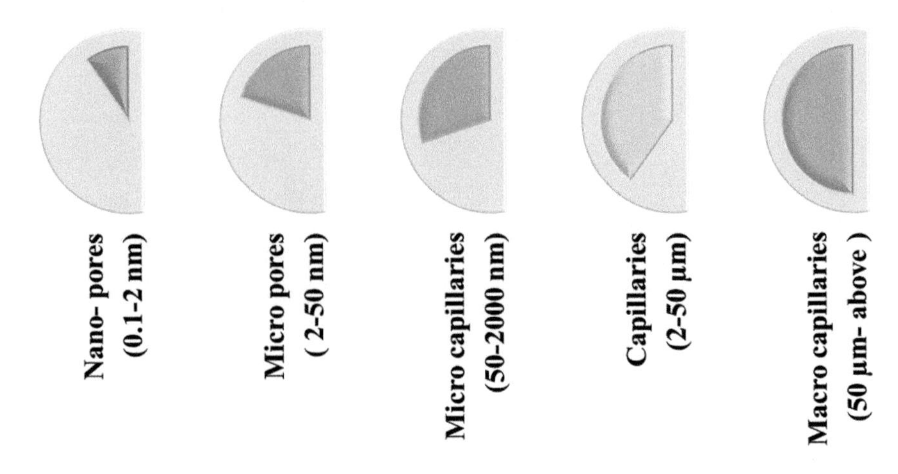

FIGURE 8.2 Classification of voids present in porous materials.

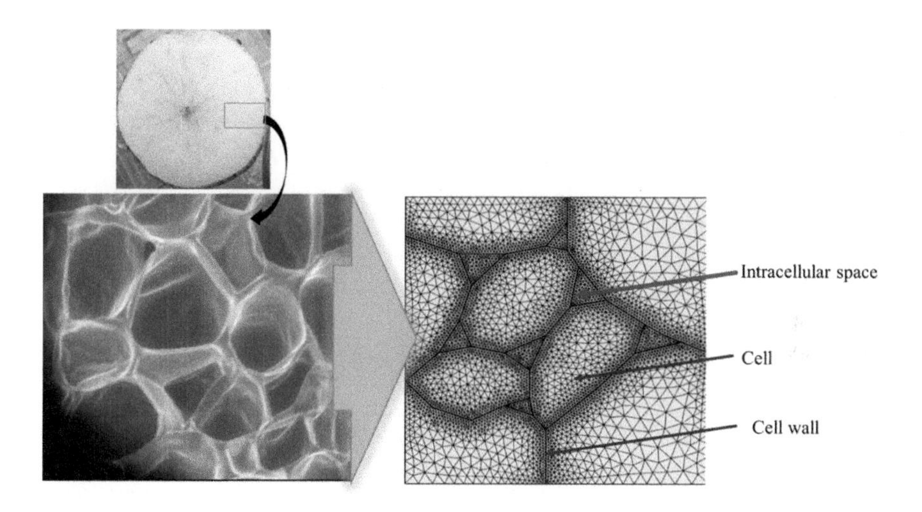

FIGURE 8.3 Micro-scale geometry and domain. (Adapted from [5].)

diameter distribution, significantly affect the macro-scale transport properties. In the micro-scale model, a different component of pores is distinguished in terms of position, shape, and orientation.

8.1.3 Transport Phenomena at the Micro-Scale

Moisture transport is greatly affected by the available size of the pore and properties of the permeable layer. Thickness and pore distribution (Figure 8.4) are the most essential dimensional parameters of moisture permeability. Therefore, modelling heat and mass transfer of food drying remains incomplete unless the transport mechanism in the corresponding length scale is considered.

In porous materials, water is distributed as free and bound water. In hygroscopic materials, such as plant-based materials, 85%–95% water remains inside

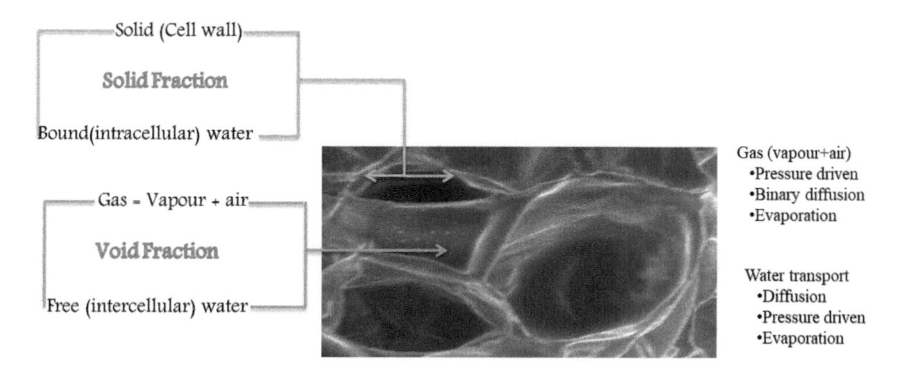

FIGURE 8.4 Micro-scale modelling domain showing different transport phenomena.

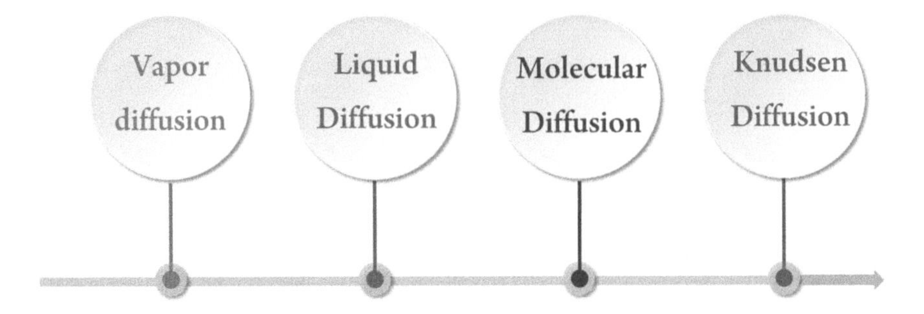

FIGURE 8.5 Diffusion mass transfer phenomena at the micro-scale domain.

the cell and is known as intracellular water, and the rest of the water remains in the intercellular spaces [7]. Several internal mass transfer phenomena are involved during drying, as discussed in Chapter 2. Different forms of diffusion mass transfer take place at the micro-scale domain. Not all of the mentioned mass transfer mechanisms, in Figure 8.5, including diffusion of vapour, liquid, and molecular and Knudsen diffusion, need to take place in a single domain.

Rather, an individual diffusion dominates in a micro-scale domain during drying. However, the accumulation of these micro-level mass transfer phenomena results in the macro-level moisture transfer coefficient.

8.1.4 Micro-Scale Modelling Approaches

Several advanced numerical methods are available to solve the micro-scale drying phenomena [8]. This section will discuss the numerical solution techniques, as shown in Figure 8.6, that have been used for solving the micro-scale drying model.

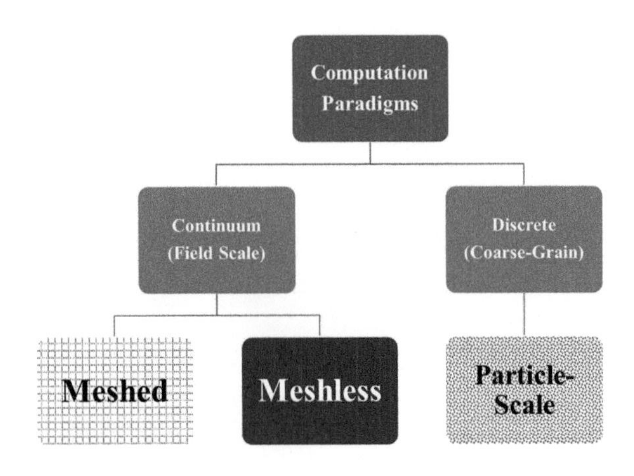

FIGURE 8.6 Computational approaches for micro-scale drying models.

8.1.4.1 Discrete Models

As the name reflects, the models are discrete rather than continuous. Discrete models are suitable for a non-continuum system. Discrete element method (DEM) is the example of the discrete model that shows a phenomenological nature. DEM is generally used to study the interaction of individual particles instead of studying the whole system. Discrete models are very capable of handling the deformation phenomena that take place during drying. As it deals with the system as a collection of particles, the simulation of the maximum number of particles is limited by computational power. Therefore, it is not a widely applied modelling approach of transport phenomena of drying.

8.1.4.2 Continuum Models

Continuum models are applicable for continuum materials. Due to the continuum nature of solid materials at the micro- and macro-level, the continuum model can successfully be applied. In the case of the fluid domain, the continuum approach works with an acceptable level of accuracy. However, there are relative advantages prevalent in mesh-considered and meshless models.

- **Meshless models:** Smoothed particle hydrodynamics (SPH) is an example of the mesh-free model. It is suitable for anisotropic materials. Due to the term "particle", many people misinterpreted SPH as a discrete model. However, SPH and other mesh-free models are completely developed for continuum solid mechanics and fluid flow-related problems. In the case of discrete models, the mesh is not concerned at all. Meshless models have been applied for micro-scale model of plant cells [9]. However, it is not as popular as mesh-based models.
- **Mesh-based models:** In mesh-based models, the domain is subdivided into numerous smaller subdomains. This subdomain is treated as volume, element, or difference in FVM, FEM, and FDM, respectively. These methods are described in Chapter 4. Out of these models, finite element method (FEM) is one of the most widely used and flexible methods. Each of the elements in the computational domain is interconnected in a certain number of nodal points. The continuous nature of the differential equation is normalized using the meshing technique.

Due to wide acceptance and simplicity in the application of FEM, we are presenting the governing equations and their associated boundary conditions in the following sections.

8.2 FEM APPROACH OF MICRO-SCALE MODELLING

Apart from the diffusion process, other moisture transport phenomena have a very small role in the transport process at the micro-level. To avoid the complexity of the model, diffusion is considered as the governing transport process. For problem formulation, please see Figure 8.3. Three variables, namely, temperature, moisture content, and airflow, are considered in this micro-scale modelling.

8.2.1 Governing Equation

Details of the governing equations associated with transport phenomena have already been presented in the previous chapter. In order to avoid repetition, we will just present the governing equations of a typical FEM model in Table 8.1.

8.2.2 Initial and Boundary Conditions

The governing equations do not consider the conditions of the interface of the cells, cell walls, and intercellular spaces; thus, the description of heat and mass transfer phenomena remains incomplete. Therefore, the mathematical expression of boundary conditions is required, as shown in Figure 8.7, to complete the overall transport phenomena. Unlike in macro-scale domain, separate boundary conditions need to be designated in the porous micro-scale domain. This is due to the non-uniform nature of boundary conditions existing for the sub-domain of the micro-scale domain. For example, a cell can be surrounded by other adjacent cells, while another cell can be surrounded by an adjacent cell and void space, as shown in Figure 8.7.

The individual line of the domain represents boundaries of the connected sub-domains. Three types of heat transfer boundary conditions can prevail in the sub-domain of micro-scale modelling, as shown in Figure 8.8.

The mass transfer boundary condition can be either mass flux or symmetry boundary conditions. As the boundary conditions were discussed in detail in the previous sections, these are not discussed.

TABLE 8.1
Governing Equations of the FEM Model

Physics			Governing Equation	Nomenclature
Heat and mass transport	Mass conversion	Water	$\rho_c \dfrac{\partial M}{\partial t} + \nabla \cdot (\rho_w v_a) = \nabla \cdot (\rho_a D_a + \nabla \rho_c D_c + \nabla \rho_w D_w)$	The subscripts used in the equations are meant as follows: a = air, w = cell wall, c = cell
		Gas	$\dfrac{\partial \rho_a}{\partial t} + \nabla(\rho_a v_a) = \nabla \cdot (\rho_a D_a)$	
	Energy conversion		$\dfrac{\partial}{\partial t}(\rho_w h_c + \rho_a h_a) = \nabla \cdot (k_w \nabla T + k_a \nabla T + h_c \rho_w D_w + h_c \rho_c D_c + h_a \rho_a D_a)$	

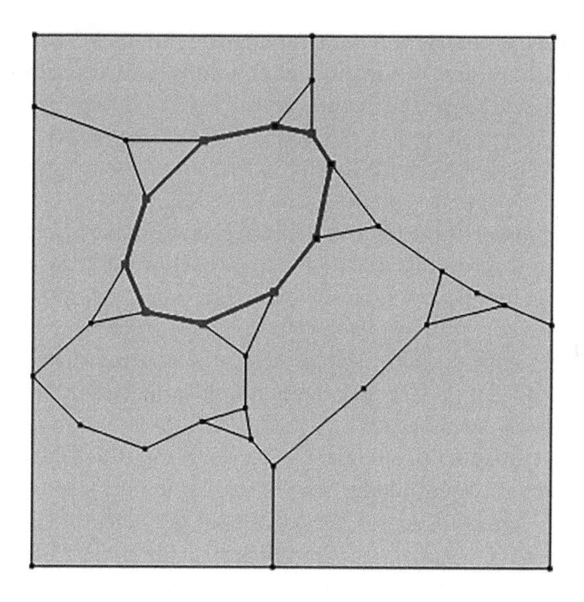

FIGURE 8.7 Boundaries of the micro-scale domain. (Adapted from [10].)

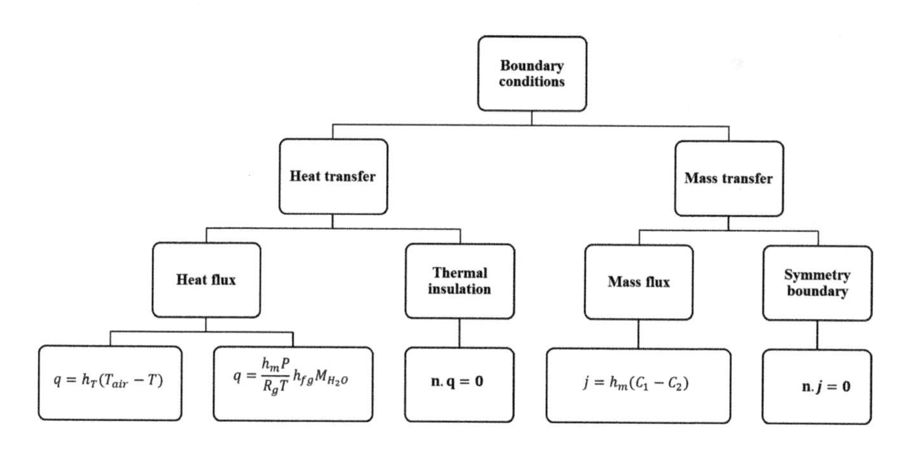

FIGURE 8.8 Boundary conditions of micro-scale drying modelling.

8.2.3 Input Parameters

Input parameters, as demonstrated in the previous chapters, mainly encompass the domain dimension and transport properties. As the micro-scale pattern varies with material types, we will not provide any detailed discussion on each of the parameters. Most of the cell-level properties are required to run a micro-level drying model for different materials. Therefore, we will present some micro-level input parameters and their measurement techniques below:

- **Micro-level dimension:** This is not a straightforward parameter as it can vary from a few nanometres to a hundred micrometres. Moreover, the porous heterogeneous nature causes further difficulties in defining the dimension of the region of interest. Special dimension measurement devices such as SEM can be used in these cases. Some typical values of micro-level domain are presented in Table 8.2.
- **Thermal properties:** Effective thermal properties have been considered for REV in macro-scale drying modelling. However, this is not a valid assumption for the micro-scale modelling approach. Special devices need to be deployed to determine thermal properties. For example, micro-level specific heat can be determined using differential scanning calorimetry (DSC), and thermal conductivity can be determined through the 3ω method [13].
- **Moisture transport properties:** Micro-level moisture transport properties are remarkably different from macro-scale ones due to the distinct structural characteristics of corresponding domains. For example, the permeability of water across the cell wall is different from that of tissue. Micro-level moisture transport properties can be determined using advanced non-destructive methods. For example, cellular-level transport properties including diffusivity and permeability can be determined experimentally using nuclear magnetic resonance (NMR) and X-ray micro-computed tomography(μCT). Micro-level moisture transfer-related properties of some typical food materials are presented in Table 8.3.

TABLE 8.2
Dimension at the Micro-Level Domain (Values Are for Cellular Food Materials)

Material	Parameter	Value (μm)	Reference
Cellular food	Cell wall thickness	4–6	[11,12]
	Cell diameter	100–500	
	Intercellular space	210–350	

TABLE 8.3
Moisture Transport-Related Properties of Some Selected Materials

Materials	Parameter	Value	Reference
Cellular food	Turgor pressure (P)	200 KPa	[14]
	Cell diffusivity (Dc)	$2.2\times10^{-9}\,m^2/s$	[15]
	Cell wall diffusivity (Dw)	$41.9\times10^{-11}\,m^2/s$	

Apart from the mentioned parameters, most of the common parameters relating to transport phenomena have already been mentioned in Chapters 6 and 7. Many parameters of the macro-level are equally valid for the micro-level domain.

8.3 TYPICAL RESULTS AND DISCUSSION

We have already discussed the typical results of macro-scale modelling in the previous chapters. Whatever simulation results can be attained at macro-level modelling, those can also be attained at micro-level modelling. Therefore, we will not present too many details in this regard except for a couple of illustrations. Temperature and moisture distribution of food materials at micro-scale are presented in the following sections.

8.3.1 Temperature Distribution

The temperature distribution over the surface of micro-scale food domain at different times is depicted in Figure 8.9. Unlike temperature distribution in the macro-level domain, heterogeneous temperature distribution prevails in the micro-scale domain. Although the initial temperature of water-rich and air-filled spaces were uniform initial temperature, due to the low specific heat of air, a faster temperature rise occurred in the intercellular spaces.

As expected, while heat propagates to the water–air interface, a different pattern of temperature distribution in the cellular and intercellular zone is observed. This is one of the distinct benefits of micro-scale modelling over macro-scale ones. Therefore, it can be claimed from Figure 8.9 that the temperature distribution within a porous material is governed by porosity, pore size distribution, thermal properties of the species, and drying air temperature.

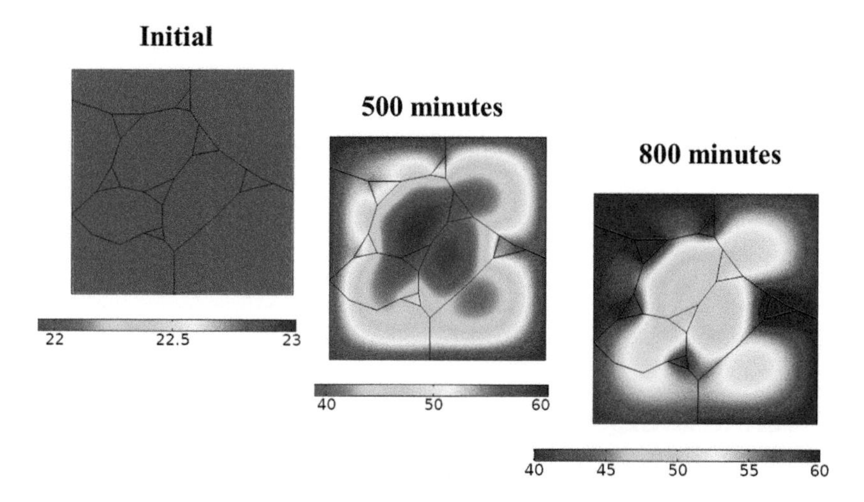

FIGURE 8.9 Temperature (°C) distribution at different times in a micro-scale cellular sample.

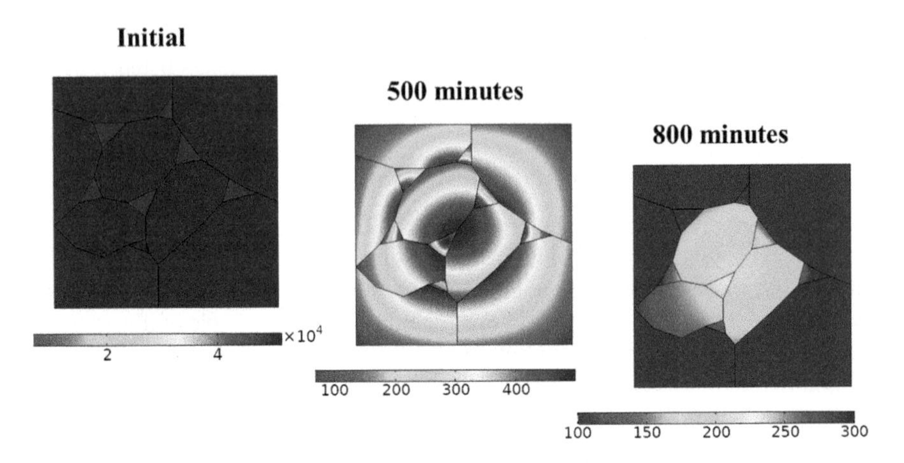

FIGURE 8.10 Moisture distribution (mole/m^3) in cellular food materials.

8.3.2 Moisture Distribution

The change in water concentration inside the cells and intercellular spaces at different times is presented in Figure 8.10. Initially, water concentration inside the cell is assumed to be uniform, and the non-uniform distribution becomes obvious during drying. Unlike macro-scale modelling, heterogeneous moisture distribution can be attained from micro-scale modelling.

Random moisture distribution in different cells and intercellular spaces can also be observed towards the later stage of the drying. These types of distribution of species are one of the most advantageous benefits of micro-scale modelling where separate domains of cell and intercellular spaces can be seen for the numerical computation of heat and mass transfer.

Moreover, as the cells have higher water content than the intercellular space, it is expected that the water transport from cell to intercellular space through the cell wall occurred [16]. Moisture transport between two cells occurs depending on the concentration and pressure gradient of the two cells. Many other moisture distribution-related parameters also vary at micro-scale from their macro-scale counterparts.

This chapter is aimed to familiarize readers with micro-scale modelling and its outcomes. The reader interested in more insight into these topics can consult numerous other sources available, such as those listed in the references.

8.4 CHALLENGES IN MICRO-SCALE MODELLING

Micro-scale modelling offers many advantages including heterogeneous moisture and temperature distribution within the sample. Despite these advantages of micro-scale model, it encounters the following challenges.

8.4.1 Domain Development

During food processing, microstructural stress due to the moisture gradient within the product leads to shrinkage. When moisture migrates from a porous material, shrinkage must take place in that material [11]. The domain required for micro-scale needs real micro-level geometry, which is still a challenging task. Especially, the real-time microstructural domain is required for micro-scale drying modelling. Non-destructive approach, including μ *CT*, scan is deployed for real microstructures for fresh samples for micro-level models; however, application of real-time micro-structure during drying is not available yet [17]. Moreover, 3D micro-scale model is not available in the literature. The 3D model is required for the sample which is anisotropic, such as plant-based food materials.

8.4.2 Boundary Conditions

Proper boundary conditions are essential for every computational modelling. Providing accurate boundary conditions in macro-scale models is much easier than for micro-scale modelling. In macro-scale modelling, the individual boundary of all subdomains including pores needs to be provided. Moreover, boundary conditions vary for different subdomains, adding more complexity.

8.4.3 Unavailability of Micro-Level Properties

It is clear from the properties and parameter section that micro-scale model requires micro-level physio-thermal and transport properties of porous materials. The required properties vary with sample types, and special experimental protocols are usually required to determine the properties. Micro-level properties are less available in literature, and determination of those is also more difficult than their macro-scale counterparts. Therefore, availability of micro-scale properties of a concerned material are essential before development and compute the micro-level model.

8.4.4 Too Much Information

Detailed information from the micro-level model regarding moisture and temperature is a great advantage of micro-scale drying modelling and is beneficial for many materials including medicines and other thermo-sensitive materials. This type of in-depth information does not add any value of the model except involving computational cost. Therefore, feasibility study of development of a micro-scale model must be carried out for the potential materials.

8.4.5 Higher Computational Cost

Simulation of micro-scale model is relatively computationally intensive than macro-scale modelling. Especially, the discrete element method incurs high computational cost. This is due to the higher number of subdomains, governing equations, and boundary conditions that cause higher computational cost. Moreover, the complexity of microstructure of porous materials computed using micro-scale models results in more complexity, and the requirement of in-depth specified properties and parameters of a micro-scale model may be hundreds of times higher than its macro-scale counterpart. Therefore, a full-scale micro-level model requires a significant amount of computation energy and capacity.

Successful implementation of micro-scale modelling can provide in-depth understanding of the associated physics during drying of porous materials. However, in real drying phenomena, one should consider not only deformation but also variable nature of drying environment; which makes the model a much more complex problem, providing a real understanding of the physics of drying phenomena. Moreover, both macro- and microscale modelling have its benefits to some appreciable extent. However, too much concession makes the macro-level model less accurate for many instances, whereas too many details of micro-scale model lead to high computational cost. The multiscale model can take the challenge to attain the benefits of models of both scales. We will give a brief description of the deformation consideration and conjugation approach in Chapters 9 and 10. Following this, Chapter 11 deals with the multiscale drying model.

REFERENCES

1. J. M. Aguilera, "Microstructure and food product engineering," *Food Technol.*, vol. 54, no. 11, pp. 56–65, 2000.
2. M. U. H. Joardder, C. Kumar, and M. A. Karim, "Multiphase transfer model for intermittent microwave-convective drying of food: Considering shrinkage and pore evolution," *Int. J. Multiph. Flow*, vol. 95, pp. 101–119, 2017.
3. C. Kumar, M. U. H. Joardder, T. W. Farrell, and M. A. Karim, "Investigation of intermittent microwave convective drying (IMCD) of food materials by a coupled 3D electromagnetics and multiphase model," *Dry. Technol.*, vol. 36, no. 6, pp. 736–750, 2018.
4. J. M. Aguilera, "Why food microstructure?," *J. Food Eng.*, vol. 67, no. 1–2, pp. 3–11, 2005.
5. M. M. Rahman, M. U. H. Joardder, and A. Karim, "Microscale geometrical model of fruit tissue for simulating cellular level changes during drying," in *The 20th International Drying Symposium (IDS 2016)*, Gifu, Japan, 7–10 August 2016, pp. 1–6.
6. M. K. Abera et al., "3D virtual pome fruit tissue generation based on cell growth modelling," *Food Bioprocess Technol.*, vol. 7, no. 2, pp. 542–555, 2014.
7. M. U. H. Joardder, M. Mourshed, and M. H. Masud, *State of Bound Water: Measurement and Significance in Food Processing*. Springer, 2019.
8. M. M. Rahman, M. U. H. Joardder, M. I. H. Khan, N. D. Pham, and M. A. Karim, "Multi-scale model of food drying: Current status and challenges," *Crit. Rev. Food Sci. Nutr.*, vol. 58, no. 5, pp. 858–876, 2018.
9. H. C. P. Karunasena, W. Senadeera, R. J. Brown, and Y. Gu, "A particle based model to simulate microscale morphological changes of plant tissues during drying," *Soft Matter*, vol. 10, no. 29, pp. 5249–5268, 2014.
10. M. M. Rahman, C. Kumar, M. U. H. Joardder, and M. A. Karim, "A micro-level transport model for plant-based food materials during drying," *Chem. Eng. Sci.*, vol. 187, pp. 1–15, 2018.

11. M. U. H. Joardder, R. J. Brown, C. Kumar, and M. A. Karim, "Effect of cell wall properties on porosity and shrinkage of dried apple," *Int. J. Food Prop.*, vol. 18, no. 10, pp. 2327–2337, 2015.
12. M. U. H. Joardder, A. Karim, C. Kumar, and R. J. Brown, *Porosity: Establishing the Relationship between Drying Parameters and Dried Food Quality*. Springer, 2015.
13. B. K. Park et al., "Thermal conductivity of biological cells at cellular level and correlation with disease state," *J. Appl. Phys.*, vol. 119, no. 22, p. 224701, 2016.
14. H. C. P. Karunasena, Y. T. Gu, R. J. Brown, and W. Senadeera, "Numerical investigation of plant tissue porosity and its influence on cellular level shrinkage during drying," *Biosyst. Eng.*, vol. 132, pp. 71–87, 2015.
15. S. W. Fanta et al., "Microscale modelling of coupled water transport and mechanical deformation of fruit tissue during dehydration," *J. Food Eng.*, vol. 124, pp. 86–96, 2014.
16. M. U. H. Joardder and M. A. Karim, "Development of a porosity prediction model based on shrinkage velocity and glass transition temperature," *Dry. Technol.*, pp. 1–17, 2019.
17. M. M. Rahman, M. U. H. Joardder, and A. Karim, "Non-destructive investigation of cellular level moisture distribution and morphological changes during drying of a plant-based food material," *Biosyst. Eng.*, vol. 169, pp. 126–138, 2018.

9 CFD Modelling of Drying Phenomena

9.1 INTRODUCTION

For most of the drying models, constant heat and mass transfer coefficients (HMTCs) are considered for the entire sample during drying [1]. This assumption causes erroneous results as HMTC continuously changes for the entire sample [2]. Conjugate HMT and fluid flow can solve this critical problem successfully. In the previous chapters, we discussed the sample with constant HMTC. In other words, ambient temperature and moisture concentration are considered constant throughout the drying environment for the entire drying time, as shown in Figure 9.1.

However, the drying environment varies both spatially and temporally, and causes non-uniform boundary conditions [3]. In the following section, we will discuss the additional concepts that are important in developing conjugate modelling.

9.1.1 Fluid Flow in the Drying

Fluid flow is associated in many unit operations including conventional drying. While modelling the sample domain-associated HMT, the effect of fluid flow is considered as a boundary condition. There is a vital need for consideration of fluid dynamics in the drying model in order to ensure sufficient accuracy in

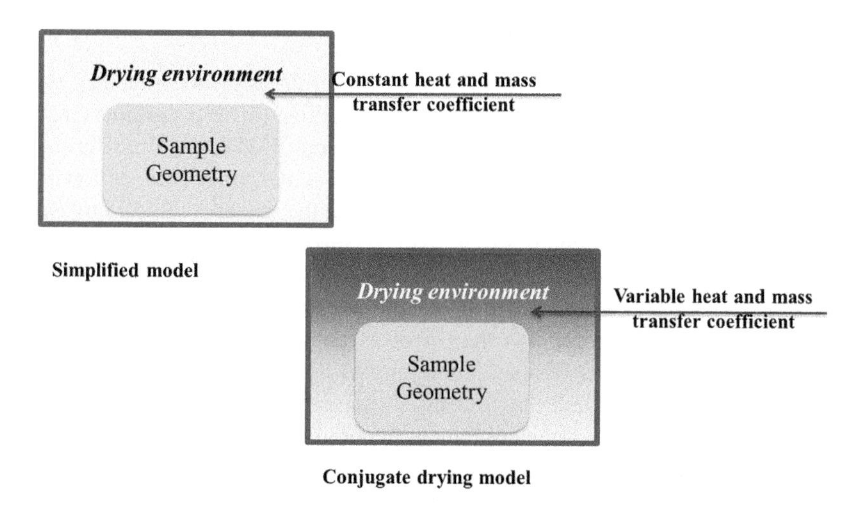

FIGURE 9.1 Simplified and real drying environmental conditions.

DOI: 10.1201/9780429461040-9

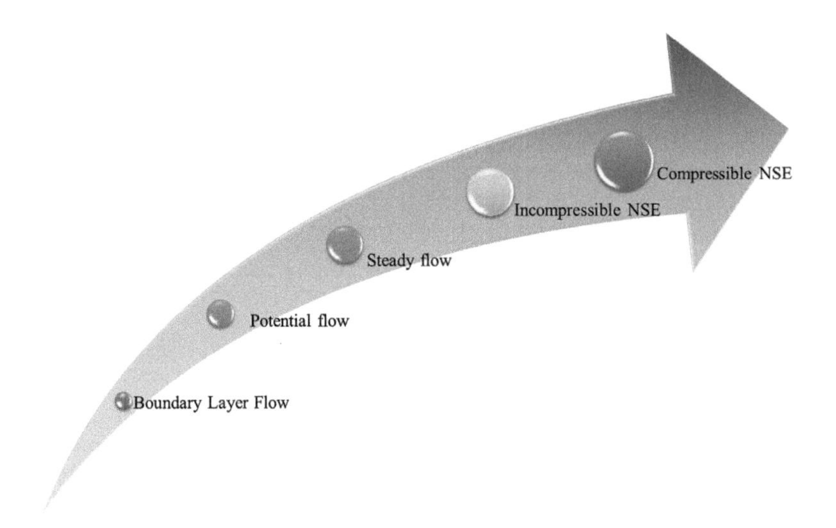

FIGURE 9.2 Types of common fluid flows.

simulation. Drying techniques that involve airflow significantly rely on HMTC [4]. Distribution of air velocity and pressure is substantially affected by sample geometry, surface condition, and flow passage configuration [5]. Depending on the nature of fluid and flow pattern, several categories of fluid flow can be observed, as shown in Figure 9.2.

The complexity of fluid flow varies in different fluid flows. The simplest flow among those mentioned in Figure 9.2 is the boundary layer flow and the most complex one is the compressible Navier–Strokes' flow. Solution technique and computational energy significantly vary with the variation of the fluid flow type.

9.1.2 Computational Fluid Dynamics

Computational fluid dynamics (CFD) is a modelling approach that has been applied widely in different unit operation modelling [6]. For instance, many of the food processors are involved in simultaneous HMT and fluid flow [7]. Understanding of the CFD modelling process is vital in drying. This section does not attempt to describe the details of CFD; instead, the reader is referred to available resources of fluid mechanics [8,9]. Among several advantages of incorporation of CFD in drying model, the quantification of spatial and temporal HMTC and observation of the effect of drying conditions are the main ones. Eventually, modelling of the drying process that involves fluid flow such as spray drying and convective drying is possible. Numerous parametric analysis of associated multi-physics including fluid flow during drying can be invesigated. Especially, the precise heat and mass transfer boundary conditions depend on the successful modelling of fluid flow.

There are many options for solving fluid flow-related problems available, and this completely depends on the nature of fluid flow [10]. Distribution of air is a complex phenomenon, and there is no absolute solution to Navier–Stroke's (N-S) equation. Especially, the involvement of fluid viscosity and turbulence results in significant complexity. Several assumptions offer quite a simplification of the Navier–Stokes equation solution. For instance, an inviscid fluid can be the simplest possible flow, whereas, neglecting turbulence, provides a laminar flow that is significantly simpler than turbulent flow. Among the turbulence models, Reynolds-averaged Navier–Stokes (RANS) equation based CFD model is applied extensively in drying-related fluid flow modelling. A very brief summary of the popular RANS turbulence model is presented in Table 9.1.

TABLE 9.1
RANS Turbulence Model Descriptions, Behaviour, and Usage

Model	Feature	Application	Computation energy (1–7)
Spalart-Allmaras	Designed specifically for wall-bounded flows on a fine near-wall mesh. Performs poorly for 3D flows and free shear flows	Suitable for relatively less complex external or internal flows and boundary layer flows.	1
Standard k-ε	Valid for fully turbulent flows. Flows involving severe pressure gradient and separation cannot be dealt with using this model	Widely used model in many turbulent flow problems.	2
RNG k-ε	Additional options incorporated in the standard k-ε model in predicting swirling	Suitable for complex shear flows involving moderate swirl and vortices.	3
Realizable k-ε	A variant of the standard k-ε model. Offers largely the same benefits and has similar applications as RNG.	Possibly more accurate and easier to converge than RNG.	4
Standard k-ω	Shows better performance for wall-bounded flows.	Well suited for free shear and low Reynolds number flows.	5
SST k-ω	Equally applicable as standard k-ω.	Dependency on wall distance makes this less suitable for free shear flows.	6
Reynolds Stress	Reynolds stresses are solved directly using transport equations, avoiding isotropic viscosity assumption of other models. Physically the soundest RANS model.	Suitable for complex 3D flows with strong streamline curvature, strong swirl/rotation.	7

Sources: (Compiled from Different Sources Including [10,11])

9.1.3 Conjugate Drying Model

Conjugation of fluid flow in simultaneous heat and mass transfer models makes drying models more realistic. The conjugated model provides more accurate prediction where fluid flow is involved along with heat and mass transfer in a drying system. As heat and mass transfer affect each other, fluid flow also influences HMT phenomenon, as demonstrated in Figure 9.3.

Coupling of fluid flow, heat transfer, and mass transfer is a quite complex task as the thermo-fluid properties of both sample and air vary significantly. To simplify this complex problem, a set of assumptions need to be considered.

9.1.4 Assumptions

- Initial temperature and moisture concentration are uniform throughout the drying environment
- Initial moisture concentration is uniform throughout the sample
- The sample is relatively smaller than the drying system and does not cause a vortex
- The sample surface is regular in shape and effect of surface friction can be disregarded
- Moist air is considered incompressible and possesses temperature-dependent properties
- The frictional heat dissipation in the moist air is neglected
- All forms of deformation of the sample are neglected

9.2 CFD-COUPLED HEAT AND MASS TRANSFER MODEL

A conjugated drying model couples a CFD model along with the HMT model. The contribution of airflow around the sample in HMT during drying can be determined using boundary conditions. In this way, the real-time HMTC can be calculated precisely.

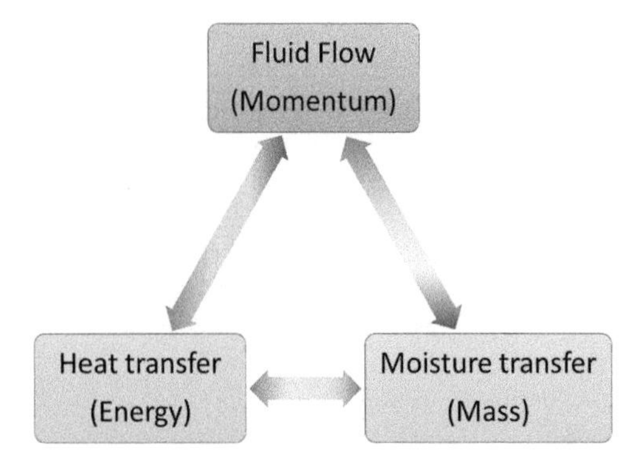

FIGURE 9.3 Interrelationship of fluid flow, heat transfer, and mass transfer in a conjugated drying model.

9.2.1 Governing Equations

Along with energy conservation (heat transfer), mass conservation (mass transfer), and ideal gas law, governing equations about fluid flow need to be considered in the conjugate model. Based on Reynolds number, first, the flow is classified as laminar or turbulent.

As we have already discussed the energy and mass conservation equations in the foregoing chapters, here we will only discuss fluid flow-related governing equations. Turbulent flow can be solved using additional turbulent terms. Reynolds averaged Navier–Stokes (RANS), detached eddy simulation (DES), direct numerical simulation (DNS), and large eddy simulation (LES) are the common solution approaches for turbulent fluid flow. Details of the sub-classifications of these modelling approaches, as shown in Figure 9.4, are available in literature, which is beyond the scope of this chapter.

To obtain air distribution, the mass conservation equation (also known as the continuity equation) is required first.

$$\frac{\partial \rho}{\partial t} + \rho \nabla \cdot (u) = 0 \tag{9.1}$$

To conserve momentum accompanied by the fluid flow, the Navier–Stokes equation is used where inertial forces, pressure forces, viscous forces, and any external forces applied to the fluid are taken into account.

$$\rho \frac{\partial u}{\partial t} + \rho (u \cdot \nabla) u = \nabla \cdot \left[-\nabla p + \mu \nabla \cdot \left(\nabla u + (\nabla u)^T \right) - \frac{2}{3} \mu \nabla (\nabla \cdot u) \right] + \text{Turbulent terms} \tag{9.2}$$

Without the turbulent terms, the abovementioned equation is used for laminar flow.

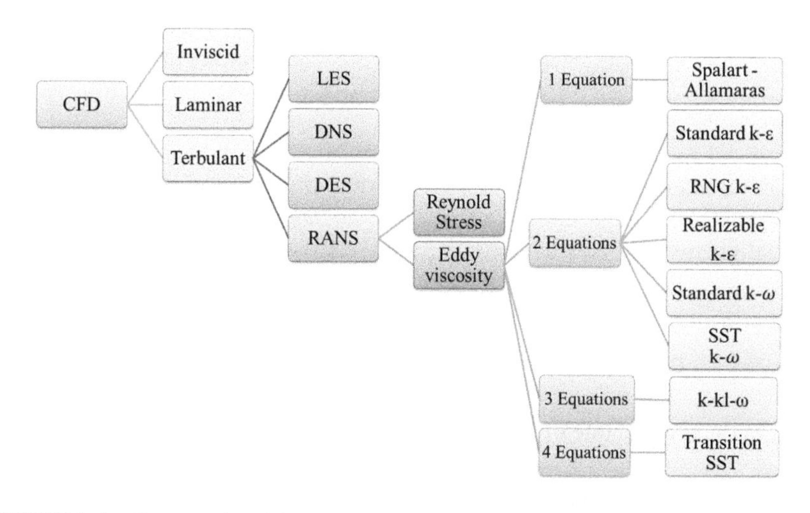

FIGURE 9.4 Computational fluid dynamic solution approaches.

9.2.1.1 Turbulent Model

Additional equation(s) is/are required along with the Navier–Stokes equation to solve the turbulent problem in order to consider turbulence in fluid flow.

9.2.1.1.1 $k-\varepsilon$ Model

$k-\varepsilon$ model takes into account two turbulent terms k and ε. In this model, k refers to the total energy content of the fluctuation of the fluid, while ε denotes the rate at which kinetic energy dissipates during turbulent flow [12]. Transfer of turbulent kinetic energy (k) can be expressed as follows:

$$\underbrace{\frac{\partial(\rho k)}{\partial t}}_{Transient}+\underbrace{\nabla\cdot(\rho Uk)}_{Convection}=\underbrace{\nabla\cdot\left[\left(\mu+\frac{\mu_t}{\sigma_k}\right)\nabla k\right]}_{Diffusion}+\underbrace{P_k-\rho\varepsilon}_{Sources+\mathrm{Sin}\,ks} \tag{9.3}$$

Besides, the transfer of turbulent energy dissipation rate (ε) can be expressed by the following equation:

$$\underbrace{\frac{\partial(\rho\varepsilon)}{\partial t}}_{\text{Transient}}+\underbrace{\nabla\cdot(\rho U\varepsilon)}_{\text{Convection}}=\underbrace{\nabla\cdot\left[\left(\mu+\frac{\mu_t}{\sigma_k}\right)\nabla_\varepsilon\right]}_{\text{Diffusion}}+\underbrace{C_1\frac{\varepsilon}{k}(P_k+C_3P_b)-C_2\rho\frac{\varepsilon^2}{k}}_{\text{Sources+Sinks}} \tag{9.4}$$

Once the transport equations for k and ε. have been solved, the turbulent term (eddy viscosity term) μ_T can be calculated simply by substituting these in the following expression:

$$\mu=C_\mu\frac{\rho k^2}{\varepsilon} \tag{9.5}$$

If $k-\omega$ is the preferred turbulence model, then Section 9.2.1.1.1 should be skipped and only the following related equation of $k-\varepsilon$ model needs to be applied.

9.2.1.1.2 $k-\omega$ Turbulence Model

This model is one of the popular RANS models for turbulent flow [13,14]. To predict turbulence, this model uses two partial differential equation (PDEs) for two variables k (kinetic energy) and ω (specific rate dissipation of k into internal thermal energy). Turbulence kinetic energy can be expressed by the following PDE:

$$\rho\frac{\partial k}{\partial t}+\rho(u\cdot\nabla)k=\nabla\cdot\left[(\mu+\mu_T\sigma_k^*)\nabla k\right]+P_k-\beta_0^*\rho\omega k \tag{9.6}$$

whereas the specific rate of kinetic energy can be written as the following PDE:

$$\rho\frac{\partial\omega}{\partial t}+\rho(u\cdot\nabla)\omega=\nabla\cdot\left[(\mu+\mu_T\sigma_\omega)\nabla\omega\right]+\alpha\frac{\omega}{k}P_k-\rho\beta_0\omega^2 \tag{9.7}$$

Turbulent term (eddy viscosity term) μ_T can be calculated by substituting the value of variables k and ε in the following expression:

$$\mu_T = \rho \frac{k}{\omega} \tag{9.8}$$

9.2.2 Boundary Conditions

In addition to the boundary conditions of heat and mass transfer, separate boundary conditions for fluid flow need to be incorporated in the model.Boundary conditions for the inlet, outlet, and two walls exist in a typical fluid domain in conjugate drying models as shown in Figure 9.5.

- **No-slip condition** is the most common boundary condition apart from the inlet and outlet of the fluid domain. No-slip refers that there is no relative velocity between a fluid and its adjacent solid wall. In other words, the velocity of the fluid at the fluid–solid interface is equal to the velocity of the wall.

  Therefore, wall boundary conditions can be expresses as $u = u_w$. If the wall velocity is zero, the boundary condition is $u = 0$.
- **Inlet boundary condition** needs to be placed at the boundary where the fluid enters the drying system. Velocity- and pressure-specified boundary conditions are the common conditions of inlet boundary condition. Moreover, temperature and turbulence characteristics of incoming air flow must be incorporated while solving turbulence equations. Sometimes, the mass flow boundary condition is applied at the inlet boundary condition.

  Velocity boundary condition $u = -U_0 n$.
- **Outlet boundary condition** needs to be placed at the boundary where the fluid exits from the drying system. Like for inlet boundary conditions, pressure-specified and velocity-specified boundary conditions are applied in outlet boundary. However, atmospheric pressure is applied in many cases when the fluid exits towards the surroundings of the dryer.

  Outlet boundary condition can also be written as $u = -U_0 n$.

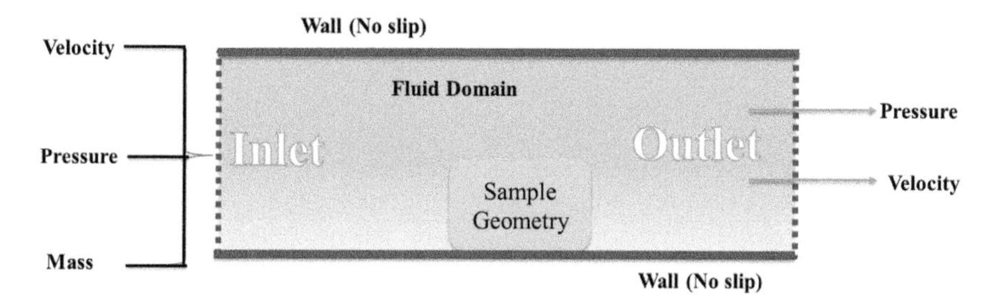

FIGURE 9.5 Boundary conditions of the fluid domain.

TABLE 9.2
Thermo-Fluid Properties of Air (Values Are at 293.15 K and 1 atm)

Description	Symbol	Value
The molecular weight of air	M_a	0.028 kg/mol
Viscosity	μ_a	1.81×10^{-5} kg/(m×s)
Thermal conductivity	$K_a{}^a$	0.025 W/(m×K)
Heat capacity	Cp_a	1.006×10^3 J/(kg×K)
Density	ρ_a	1.205 kg/m^3
Free stream velocity	u_0	0.5 (m/s)
Ambient pressure	P_0	1.0133×10^5 Pa

9.2.3 Input Parameters

Input parameters as demonstrated in the previous chapters mainly encompass the domain dimension and heat and mass transport properties. The additional properties required are the instantaneous properties of moist air. Therefore, we present the required properties of moist air in Table 9.2.

Apart from the mentioned parameters, an individual model of turbulent terms needs several values of its associated constant terms. These constant parameters are available in the literature.

9.3 TYPICAL RESULTS AND DISCUSSION

In the previous chapter, the temporal and spatial distribution of temperature has been presented along with moisture distribution. In those cases, thermo-fluid properties of drying air have been considered constant. Using conjugation of fluid flow in the previous model, instantaneous conditions of the moist air of the drying domain can easily be observed along with the drying kinetics of dried material. In the following section, moisture, temperature, humidity, and velocity distributions in typical drying systems and samples are discussed briefly.

9.3.1 Temperature Distribution

The temperature distribution in the fluid domain and the sample at different times is depicted in Figure 9.6. Like for temperature distribution in dried sample domain, temperature gradient prevails in the fluid domain. Separate uniform initial temperatures are considered for sample and drying domain.

As expected, while heat flows over the sample, heat is absorbed by the sample, with the resulting increase of temperature. On the other hand, evaporation of water causes a decrease in the air temperature which is adjacent to the sample. The gradient of the temperature of air adjacent to the sample exists in the early part of drying. However, it proceeds to the thermal equilibrium at the later stage of drying.

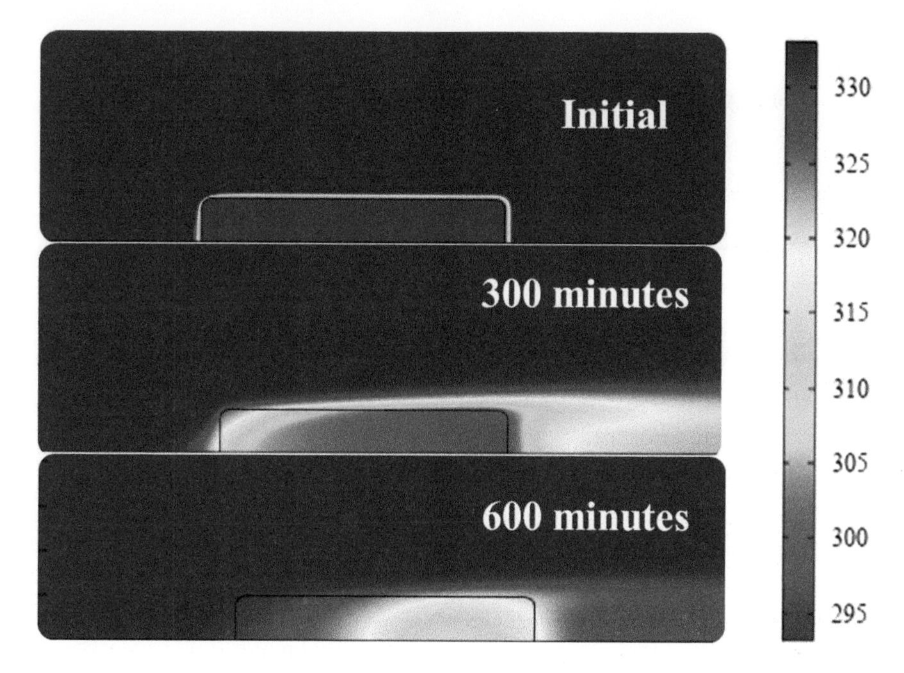

FIGURE 9.6 Temperature (K) distributions at different times in a micro-scale cellular sample.

9.3.2 Liquid Water Content

The change in liquid water concentration at different times is presented in Figure 9.7. Initially, the water concentration in the sample is assumed uniform but the non-uniform distribution become obvious during drying.

Water is removed at different rates at different locations of the sample depending on the flow of air. Water is removed faster at the leading edge where the air strikes first. The tail edges experience less water removal due to higher moisture content of the air and lower temperature of the air.

9.3.3 Relative Humidity

The change in water vapour concentration of drying air and sample at different times is presented in Figure 9.7. Initially, vapour concentration inside the sample is assumed saturated and the non-uniform distribution becomes obvious during drying.

It is clear from Figure 9.8 that the surrounding air becomes enriched in water vapour with the progress of drying. The relative humidity of the air that is adjacent to the sample is higher than that of the free stream. This increment of RH is due to the additional moisture from the sample in the drying air.

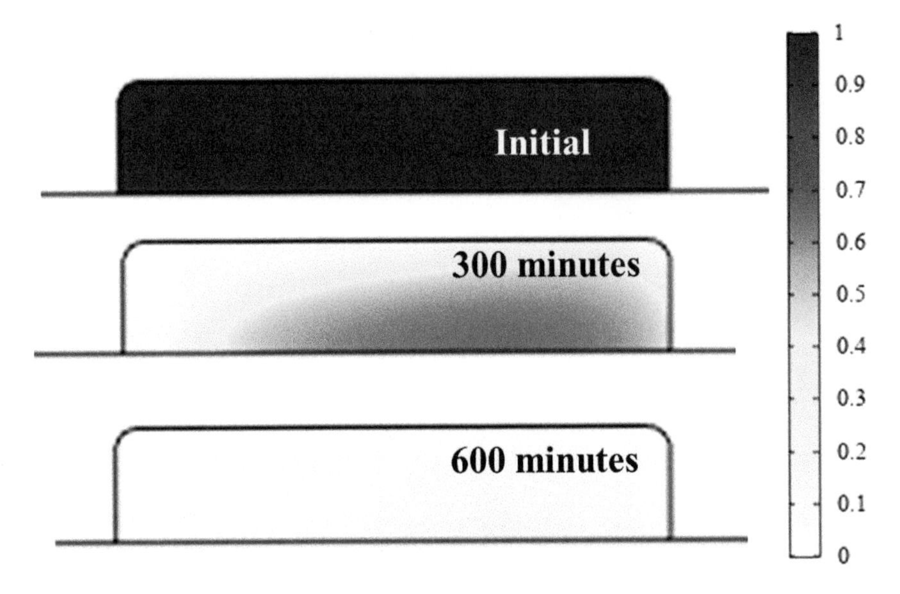

FIGURE 9.7 Moisture distribution (mole/m^3) in cellular food materials.

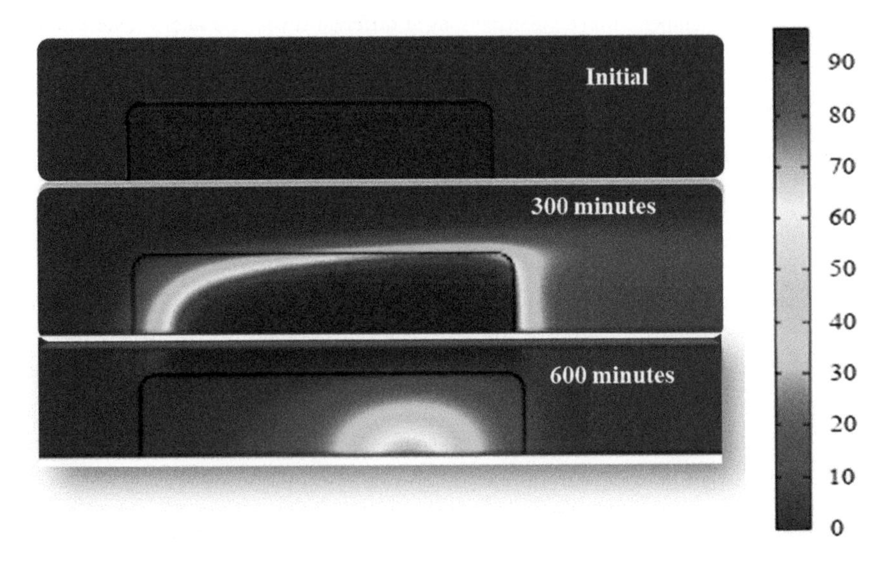

FIGURE 9.8 Relative humidity in the drying air.

9.3.4 Air Velocity

Air velocity varies significantly throughout the dryer. The distribution of air velocity inside the dryer is depicted in Figure 9.9. Air velocity distribution depends on many things, including the shape and size of dryer and sample.

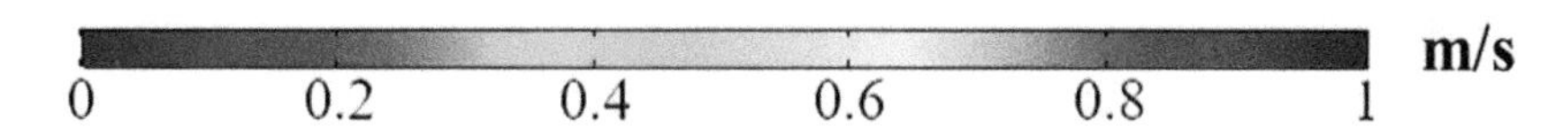

FIGURE 9.9 Air distribution in the drying system.

Due to the no-slip condition, the velocity of air adjacent to both walls is close to zero. The velocity is maximum at the centre of the dryer at the current shape. This is not always the case for other types of dryers. Air velocity close to the sample is also significantly affected due to the stagnant and boundary separation of the airflow.

Fluid flow and heat and mass transfer can be visualized in successful modelling of a conjugated drying model. Successful application of the conjugate model can offer a great tool for optimizing drying conditions for an efficient dryer. Although remarkable advancements in the application of CFD in drying have been made, lack of details on the physicochemical properties of samples most often make conjugate modelling difficult in real life.

REFERENCES

1. T. Defraeye, B. Blocken, and J. Carmeliet, "Analysis of convective heat and mass transfer coefficients for convective drying of a porous flat plate by conjugate modelling," *Int. J. Heat Mass Transf.*, vol. 55, no. 1–3, pp. 112–124, 2012.
2. M. I. H. Khan, Z. Welsh, Y. Gu, M. A. Karim, & B. Bhandari, "Modelling of simultaneous heat and mass transfer considering the spatial distribution of air velocity during intermittent microwave convective drying," *Int. J. Heat Mass Trans.*, vol. 153, p. 119668, 2020
3. N. Duc Pham, M. I. H. Khan, M. U. H. Joardder, A. M. N. Abesinghe, and M. A. Karim, "Quality of plant-based food materials and its prediction during intermittent drying," *Crit. Rev. Food Sci. Nutr.*, vol. 59, no. 8, pp. 1197–1211, 2019
4. M. H. Masud, R. Ahamed, M. U. H. Joardder, and M. Hasan, "Mathematical model of heat transfer and feasibility test of improved cooking stoves in Bangladesh," *Int. J. Ambient Energy*, vol. 40, no. 3, pp. 317–328, 2019.
5. M. M. Rahman, M. U. Joardder, M. I. H. Khan, N. D. Pham, and M. A. Karim, "Multi-scale model of food drying: Current status and challenges," *Crit. Rev. Food Sci. Nutr.*, vol. 58, no. 5, pp. 858–876, 2018
6. J. Ostanek and K. Ileleji, "Conjugate heat and mass transfer model for predicting thin-layer drying uniformity in a compact, crossflow dehydrator," *Dry. Technol.*, vol. 38, no. 5–6, pp. 775–792, 2020.

7. M. U. H. Joardder, C. Kumar, and M. A. Karim, "Multiphase transfer model for intermittent microwave-convective drying of food: Considering shrinkage and pore evolution," *Int. J. Multiph. Flow*, vol. 95, pp. 101–119, 2017.
8. C.-Y. Chow, *An Introduction to Computational Fluid Mechanics.* John Wiley & Sons, 1979.
9. A. C. Yunus, *Fluid Mechanics: Fundamentals and Applications (Si Units).* Tata McGraw Hill Education Private Limited, 2010.
10. H. Lomax, T. H. Pulliam, and D. W. Zingg, *Fundamentals of Computational Fluid Dynamics.* Springer Science & Business Media, 2013.
11. S. B. Pope, *Turbulent Flows.* IOP Publishing, 2001.
12. W. P. Jones and B. E. Launder, "The prediction of laminarization with a two-equation model of turbulence," *Int. J. Heat Mass Transf.*, vol. 15, no. 2, pp. 301–314, 1972.
13. D. C. Wilcox, "Formulation of the k–ω turbulence model revisited," *AIAA J.*, vol. 46, no. 11, pp. 2823–2838, 2008.
14. D. C. Wilcox, *Turbulence Modeling for CFD*, vol. 2. DCW Industries, 1998.

10 Modelling of Deformation During Drying

10.1 INTRODUCTION

Deformation is an indispensable physical phenomenon observed in materials during different dehydration processes. This change affects heat and mass transfer process and many quality attributes of dried products. Many mathematical models ranging from empirical to classical models are proposed in the literature for predicting deformation during drying of materials. Deformation is a generic term that means either shrinkage or expansion as shown in Figure 10.1.

The sample may encounter either shrinkage or expansion depending on the drying conditions and material properties. A sample may also go through both shrinkage and expansion at different stages of drying. Due to the heterogeneous hierarchical structure of porous materials including food, many of the

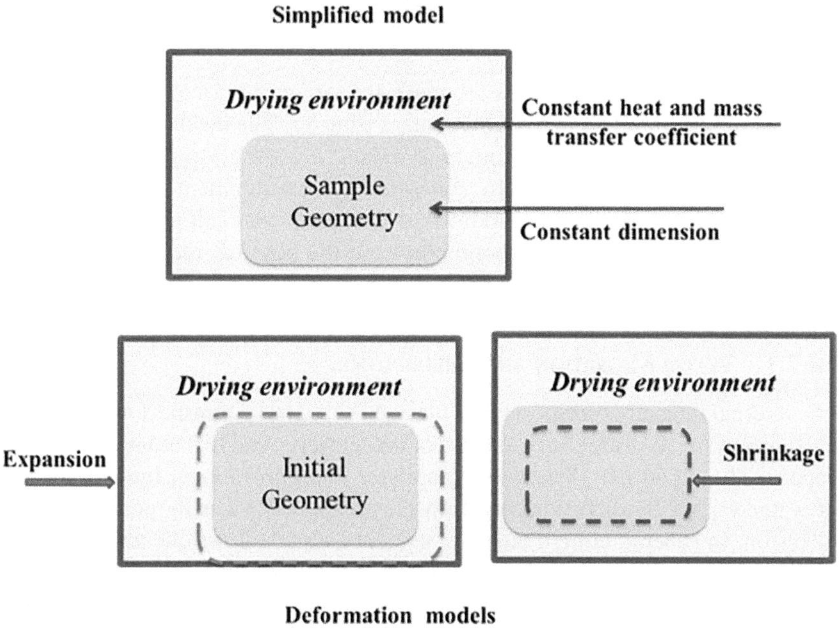

FIGURE 10.1 Deformation phenomena during drying.

DOI: 10.1201/9780429461040-10

materials are highly shrinkable in the condition of drying. Shrinkage of dried products has many consequences including surface cracking and reduction of rehydration capability [1–3]. Shrinkage also has a great impact on the mechanical and textural properties of dried materials [4]. Most importantly, shrinkage is an important factor that significantly affects the drying rate as well as drying kinetics.

Taken these concerns into consideration, deformation should not be neglected while predicting actual heat and mass transfer during drying [5]. As the deformation of porous materials depends on many factors, including material characteristics, microstructure, mechanical properties, and process conditions, prediction of deformation is not a straightforward process.

Several researchers proposed mathematical expressions to predict the deformation of food materials as a function of the moisture content. These models can be grouped into three categories: (i) theoretical models that are built based on the understanding of the fundamental physics and mechanisms that may be involved in deformation [6–8], (ii) empirical models that are built by fitting the model's parameters to the experimental data [9–11], and (iii) semi-empirical, meaning a blender of empirical and theoretical nature.

In this chapter, we will discuss the modelling approach of deformation during drying phenomena. The coupling of the transport model and the deformation model is not described in this chapter in great detail due to the limited scope of the issue for this current book.

10.2 FACTORS ASSOCIATED WITH DEFORMATION DURING DRYING

During the drying of any porous products, the structural orientation is affected because of the simultaneous heat and mass transfer. The mechanisms of deformation in porous materials throughout drying are very complicated because of the structural complexity. In general, void development because of the water migration and structural mobility is a more prominent phenomenon that causes the deformation. This section lays out the physical mechanisms of the deformation.

10.2.1 Water Migration and Distribution

The internal moisture transport takes place utilizing three transport mechanisms: (i) transport of free water, (ii) transport of bound water, and (iii) transport of water vapour. The liquid may flow due to capillary and gravitational force, transport of water vapour by diffusion, and transport of bound water by desorption and diffusion. In general, at high moisture contents, liquids flows due to dominating capillary forces. With decreasing moisture content, the amount of liquid in the pores also decreases, and a gas phase is built up, causing a decrease in the liquid permeability. Gradually, the mass transfer is taken over by vapour diffusion in a porous structure, with increasing vapour diffusion [12].

Deformation that takes place during drying is caused by the void space left by the migrated water. However, shrinkage of the solid matrix of the sample compensates for the void volume. Deformation can be expressed by shrinkage, collapse, or expansion of porous materials during drying. The degree of these structural changes depends on the distribution of water within the porous material and the structural rigidity of the sample material.

In general, porous materials undergoing drying are subjected to stresses, leading to shape modification and deformation. These stresses are developed due to concurrent thermal and moisture gradients. The stress generated by moisture gradients takes place during the entire period of the drying process, while the thermal stress is stronger at the earlier stage of the drying. Therefore, the stress generated by the moisture gradient is the dominant cause of the shrinkage of plant tissue during drying.

Due to the diverse nature of different plant tissues, it is not possible to explain deformation throughout drying using the water migration and distribution concept.

10.2.2 Structural Mobility

The collapse of cells and shrinkage occur in porous materials during the drying process. A subtle distinction exists between shrinkage and collapse; shrinkage refers to a reversible reduction in the volume of the food sample, whereas collapse represents an irreversible breakdown of the structure at either a cellular or a tissue level [13].

To explain these physicochemical changes, Slade and Levine [14] first introduced the glass transition concept in food materials during storage and drying. The glass transition temperature (T_g) is defined as the temperature at and above which an amorphous material changes from glassy state to rubbery state [15].

In the glassy state, the viscoelastic material matrix shows very high viscosity of the range of 10^{12} to 10^{13} Pa s. Effectively, the matrix behaves like a solid as it can retain the rigidity of the structure by supporting its body weight against the flow of the solid matrix due to gravity [16]. Therefore, $\Delta T = (T - T_g)$ can be treated as the driving force of structural collapse [17], where T and T_g are the temperatures of a sample undergoing drying and glass transition temperature, respectively. The effect of glass transition temperature on the rate of shrinkage can be described as follows:

a. The degree of shrinkage is related to the state of the solid matrix [18]. Rubbery state of material shows high mobility of solid matrix, whereas low mobility is caused when the sample reaches glassy state [19].
b. Shrinkage is highly dependent on the value of mobility temperature. Generally, a high rate of shrinkage is caused due to higher mobility temperature that can be observed at the early stage of drying [20,21].
c. The sample undergoes shrinkage until the mobility temperatures reach zero. Therefore, drying at higher temperature causes shrinkage even at very low moisture content [22].

Although the glass transition concept can be applied to explain or identify the deformation pattern to some extent during food processing, many experimental studies were unable to establish the relevant relationship to support this concept [17,23].

10.2.3 Phase Transition (Multiphase)

Porous materials, including plant-based food materials, encompass three phases of materials, namely, solid, liquid, and gas [24]. The void spaces are filled with multiphase species including water, vapour, and air materials. Deformation occurs with the changes of phases of liquid to gas and their associated transportation as discussed earlier.

The phase change from liquid water to vapour is an important phenomenon as the diffusion of the vapour phase is a significant mass transfer mechanism in the porous food sample. Therefore, multiphase mass transfer phenomenon occurs during drying. In many cases, vapour diffusion accounts for 10%–40% of total moisture flux in convective drying at 55–71°C [25]. In addition, vapour diffusion is faster than liquid water diffusion in porous media during drying. The crust formation effect can be observed if drying conditions allow a phase transition (liquid water to vapour) in the outer zone material, even at high drying rates [26]. Another remarkable phenomenon, crust formation occurred due to the precipitation of non-volatile compounds on the surface [27].

Evaporation rate has a substantial effect on the deformation during drying. In particular, over the time of drying under microwave energy, higher microwave power leads to increased evaporation rate; consequently, the more porous structure developed in comparison with drying with lower microwave drying. Therefore, fast evaporation of water retains the porous structure and causes less collapse of the solid matrix.

In brief, the microstructure and solid matrix architecture of porous materials vary so much that there are many exceptions of these explanations of the pore formation. Therefore, various concepts such as water-holding capacity, water distribution, glass transition temperature, the phase change of liquid water, and internal moisture transport mechanisms need to be taken into consideration to explain deformation clearly for a particular material in a selected drying condition.

10.3 MATHEMATICAL MODELS FOR DEFORMATION

In the drying process, deformation is caused due to pore formation and shrinkage of the solid matrix. The void space developed due to water migration results in pore formation or shrinkage of solid matrix compensates for this. When the volume reduction of the product is exactly equal to the volume of the removed water during drying, ideal shrinkage is found. If this occurs, it is assumed that there is no pore formation taking place in the product. On the other hand, when the volume reduction is smaller than the volume of migrated water, pore formation has taken place. To predict the deformation during drying, a large number of models have been developed.

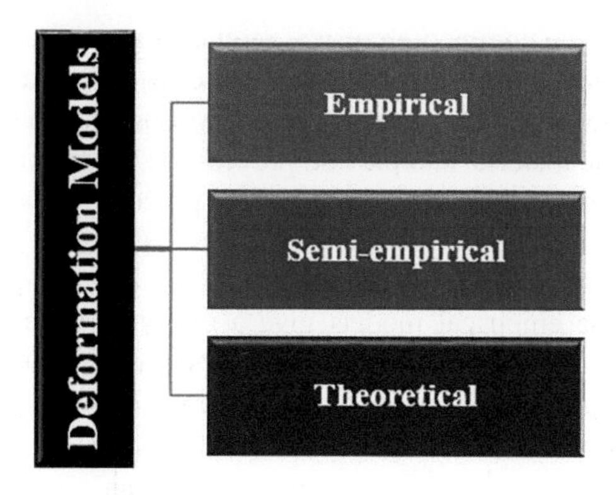

FIGURE 10.2 Classification of deformation modelling.

Deformation modelling can be classified as shown in Figure 10.2. Empirical models are developed using experimental data to fit parameters for a particular system. Methods, materials, and the processing environment restricts these models, and the fitting parameters usually have no physical significance.

There is an advanced type of the empirical model that considers physical phenomena to a lesser extent than the theoretical model; however, it is far better than pure empirical models. A comprehensive review of the literature confirmed that both product characteristics and process parameters discernibly affect deformation during drying of food materials. Therefore, any of the parameters or conditions can improve the quality of the prediction.

In contrast, theoretical modelling can provide better insights into the deformation that accompanies simultaneous heat and mass transfer during drying. However, limited attempts have been made on the theoretical modelling of deformation of fruits and vegetables due to the complexity of developing physics-based models. It is worth mentioning that deformation models are sometimes also referred to as shrinkage or porosity modelling. Therefore, the terms are interchangeably used by considering their interrelationships. In the following section, we briefly describe the mentioned types of deformation modelling.

10.3.1 Empirical Models

Most of the available studies regarding shrinkage are either experimental or empirical. These empirical models of shrinkage cannot provide a realistic understanding of the physical phenomena behind deformation during drying. The first porosity model was published in the 1950s when Kilpatrick et al. [28]. They developed a simple model taking the volumetric shrinkage of fruits and vegetables during drying into account. Most of the models found in the literature considered ideal shrinkage during drying where it is assumed that the geometrical volume reduction of the product is exactly equal to the volume of the removed water. These studies predicted shrinkage by empirical correlations as a function of water content in the form of linear [29], quadratic [30], exponential [31], and power-law equations [11,32].

In Table 10.1, some common empirical models that were developed using the correlation of the moisture content of the sample with shrinkage or porosity are presented. These models have a higher accuracy of prediction but are only applicable to the specific material sample and process the model was tested for. Thus, they cannot be used for diversified products and processes or as a tool for prediction because slight changes in process conditions and sample parameters significantly alter the value of fitting parameters.

Even though the empirical models give a reasonable prediction, they offer limited insight into the fundamental principles involved during deformation. Therefore, a clear understanding of the relationships between porosity and transport process could not be achieved.

TABLE 10.1
Summary of Empirical Model Shrinkage of Plant-Based Materials during Drying [33]

Empirical Equation	References
$\frac{V}{V_0}=\frac{1}{(1-\varepsilon)}\left[1+\frac{\rho_0(X-X_0)}{\rho_w(1+X_0)}\right]$	[10]
$\frac{V}{V_0}=\frac{1}{(1-\varepsilon)}\left[\frac{\left(\frac{\chi_{cw}}{\rho_{cw}}+\frac{\chi_{sg}}{\rho_{sg}}+\frac{\chi_{st}}{\rho_{st}}+\frac{X}{\rho_{sn}}\right)\rho_0}{(1+X_0)}\right]$	[6]
$\frac{V}{V_0}=\frac{1}{(1-\varepsilon)}\left[1+\frac{\rho_0(X-X_0)}{\rho_w(1+X_0)}-\varepsilon_0\right]$	[26]
$\frac{V}{V_0}=\frac{\rho_0}{\left(\frac{1-\varepsilon_{ex}-\varepsilon}{\sum_{i=1}^{n}\frac{M_i}{(\rho_T)_i}}\right)}\left[\frac{\rho_0}{(1+X_0)}\right]$	[34]
$\frac{V}{V_0}=\frac{1-\varepsilon_0}{1-\varepsilon}\frac{1}{\left(\frac{\rho_S}{\rho_W}\right)X+1}\left[1+X\left(\frac{X}{X_0}\frac{\rho_S}{\rho_W}\right)\right]$	[35]
$\varepsilon=1-\frac{\rho_b}{\rho_s}$	[28]
$\varepsilon=\frac{(X+1)}{(X_0+1)}\frac{\rho_{b0}}{\rho_b}$	[6]
$\varepsilon=1-\frac{\rho_{b0}\left[\rho_w+X\rho_s\right]}{\rho_w\rho_s\left[1+\beta X\right]}$	[1]

Nomenclature

$\frac{V}{V_0}=\beta=$shrinkage
V=instantaneous volume
V_0= initial shrinkage
ε=porosity
X=moisture content (dry basis)
M=mass fraction (kg/kg, total mass)
χ =volume fraction of water (volume of water/total volume)
ρ=density

Subscript

cw=cell wall material
sg=sugar
st=starch
w=water
ex=excess volume
i=component
n=number of components
s=solid
0=initial

10.3.2 Semi-Theoretical Models

Limited studies incorporated some of the physical phenomena in empirical modelling to predict the deformation of the sample during drying. Structural parameters, like the volumetric shrinkage coefficient, shrinkage-expansion coefficient, and shrinkage-collapse coefficient in these models represent physical phenomena during deformation. However, the determination of the values of these structural parameters requires experimental investigation. Therefore, these parameters are completely process dependent and make the models less theoretical or generic. Even a slight change in process variables results in a different value of these parameters [36]. In Table 10.2, two models which are semi-theoretical are demonstrated.

Researchers also attempted to develop deformation modelling using artificial neural networks (ANNs). When sufficient experimental data is available, ANN offers accurate and cost-effective methods of developing helpful relationships between the selected variables [39]. There are many types of ANNs, including multilayer perception, recurrent neural networks, and radial basis function networks.

TABLE 10.2
Semi-Theoretical Model Shrinkage of Plant-Based Materials During Drying [33]

References	Empirical Equation	Nomenclature
[37]	$\varepsilon' = \frac{\phi\alpha}{\phi\alpha + \beta'}$	$\frac{V}{V_0} = \beta =$ shrinkage V=instantaneous volume V_0=initial shrinkage ε=porosity X=moisture content (dry basis) M=mass fraction (kg/kg, total mass) ρ=density **Subscript** w=water 0=initial $\phi = \frac{\beta'(\varepsilon_0 - \varepsilon)}{\alpha(\varepsilon - 1)}, \alpha = \frac{1-M}{\rho_w}$ $\beta' = \frac{M}{\rho_m}, \varepsilon = \varepsilon_0 + (1-\varepsilon_0)\frac{\alpha}{\alpha + \beta'}$ $A = \varepsilon_0\delta(X) + X_0\beta[\phi(X)(1-\varepsilon_0) + \varepsilon_0\delta(X)]$ $B = \beta\phi(X)(1-\varepsilon_0)$ $C = 1 - \varepsilon_0 + \varepsilon_0\delta(X) + X_0\beta[\phi(X)(1-\varepsilon_0)\varepsilon_0\delta(X)]$ $D = \beta(1-\varepsilon_0)[1-\phi(X)]$
[38]	$\varepsilon(X) = \frac{A + BX}{C + DX}$	

TABLE 10.3
Porosity Prediction Model Using Hybrid Neural Networks

Inputs	Porosity Expression	Absolute Mean % Error
Temperature, water content	$\varepsilon = 0.5X_w^2 - 0.8X_w - 0.002T^2 + 0.02T - 0.15$	56.5
Temperature, water content, initial porosity	$\varepsilon = 0.5X_w^2 - 0.8X_w - 0.002T^2 + 0.02T - 0.15(1-\varepsilon_0)$	0.98
Temperature, water content, initial porosity, product type	$\varepsilon = 0.5X_w^2 - 0.8X_w - 0.002T^2 + 0.02T - 0.05(1-\varepsilon_0)F$	0.58

As deformation prediction depends on different factors like drying method, drying parameters, and material types, a large amount of experimental data for different variables affecting deformation is required to develop an ANN model.

Hussain et al. [39] developed a generic model for the porosity prediction using a hybrid neural network. In that model, 286 data were collected for four input variables: temperature, water content, initial porosity, and type of product. ANN models were found in that investigation depending on the inputs is presented in Table 10.3.

A recent work of Joardder and Karim [33] presented a new concept of the semi-theoretical model of deformation prediction. Consideration of glass transition temperature takes into account the phase change from glassy to the rubbery state during drying, whereas the moisture diffusivity takes into account the shrinkage due to moisture migration during drying. Therefore, consideration of these two parameters in the shrinkage velocity takes process parameters and material properties into account for predicting deformation during drying. The physical meaning of shrinkage velocity is simply the velocity of the outer surface of the sample during shrinkage.

$$v_s = \frac{D_{eff}\rho_w\left(T - T_g\right)}{2l\rho_p\left(T_o - T_{g0}\right)} \tag{10.1}$$

where v_s is the shrinkage velocity in ms^{-1} and l is the half-thickness of the sample.

T, T_0=instantaneous and initial temperature of the sample (K)

T_g, T_{g0}=instantaneous and initial glass transition temperature of the sample (K)

After getting the instantaneous bulk volume from the above shrinkage velocity approach, porosity of the material during drying can be predicted by the following equation:

$$\varepsilon = 1 - \frac{V_w + V_s}{\pi\left(r_0 - v_s t\right)^2\left(L_0 - v_s t\right)} \tag{10.2}$$

where V_w is the instantaneous volume of water and V_s is the volume of solid material of the sample that is considered constant throughout the drying process. Also, t is the time of drying, r is the instantaneous radius of the sample, and L is the instantaneous thickness of the sample. This model has been applied for selected food materials and has potential in the application of other materials under drying condition.

10.3.3 Theoretical Models

Similar to empirical models, semi-empirical models depend on the experimental results for some particular parameters. Therefore, it is essential to develop theoretical models as they have a wider insight into physical parameter changes that can be used for a wide range of processes and products (Figure 10.3).

There are many material models available in the literature. Some theoretical shrinkage models based on continuum mechanics are shown in Table 10.4.

There are many studies on the theoretical models for predicting deformation. As the mechanistic models depend on the characteristics of food materials, material characterization is crucial to develop a theoretical deformation model [44]. Hooke's law has been applied for food material, but ideally it is not valid for biological materials [45] because of its limitation when applied to large deformation. As the mechanistic models depend on the characteristics of food materials, material characterization is crucial to develop a theoretical shrinkage model [44].

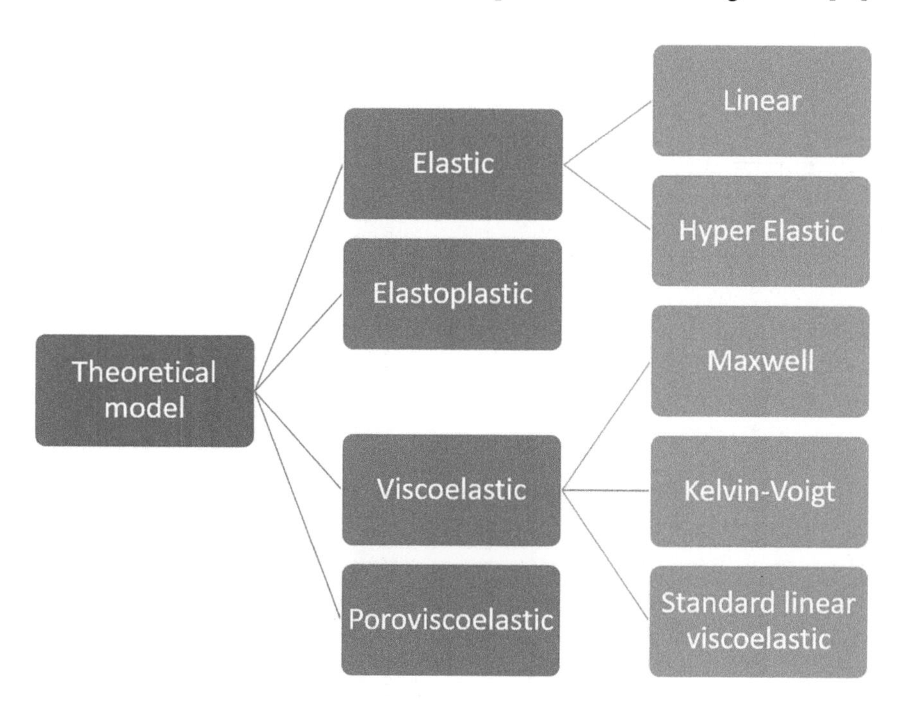

FIGURE 10.3 Theoretical deformation model classification.

TABLE 10.4
A Brief Compilation of Theoretical Material Model of the Porous Model on Which Deformation Models Can Be Achieved [40–43]

Model	Equation	Remarks
Linear elastic	$d\sigma_{ij} = \frac{E}{1+\nu}\left[d\varepsilon_{ij}^{p} + \frac{\nu}{1-2\nu}\left(d\varepsilon_{xx}^{p} + d\varepsilon_{yy}^{p} + d\varepsilon_{zz}^{p}\right)\delta_{ij}\right]$	• The strains in the material are small (linear) • Stress is proportional to the strain, (linear) • Material returns to its original shape when the loads are removed, and the unloading path is the same as the loading path (elastic) • There is no dependence on the rate of loading or straining (elastic)
Hyper-elastic model	$U_s = \frac{\mu}{2}\left(\overline{I}_1 - 3 - 2lnJ_{el}\right) + \frac{\eta}{2}\left(lnJ_{el}\right)^2$	• Nonlinear stress–strain behaviour
Viscoelastic model	$\sigma + \frac{\eta}{E}\dot{\sigma} = \eta\dot{\varepsilon}$	• Time-dependent recovery of deformation followed by load removal stress relaxation under constant deformation, and time-dependent creep rupture
Elastoplastic model	The elastoplastic stress-strain matrix is required to solve the elastoplastic model. The elastoplastic stress-strain matrix can be formulated as: $\begin{Bmatrix} d\sigma_{\gamma} \\ d\sigma_{z} \\ d\sigma_{\theta} \\ d\tau_{\gamma z} \end{Bmatrix} = \left\{ \frac{E(1-\nu)}{(1+\nu)(1-2\nu)} \begin{bmatrix} 1 & \frac{\nu}{1-\nu} & \frac{\nu}{1-\nu} & 0 \\ \frac{\nu}{1-\nu} & 1 & \frac{\nu}{1-\nu} & 0 \\ \frac{\nu}{1-\nu} & \frac{\nu}{1-\nu} & 1 & 0 \\ 0 & 0 & 0 & \frac{1-2\nu}{2(1-\nu)} \end{bmatrix} \begin{Bmatrix} d\varepsilon_{\varsigma\gamma} \\ d\varepsilon_{\varsigma z} \\ d\varepsilon_{\varsigma\theta} \\ d\gamma_{\varsigma\gamma z} \end{Bmatrix} \right\}$	• Strain displacement is proportional to the shrinkage coefficient
Poro-viscoelastic model	For linear viscoelastic system, the effective solid stress tensor becomes, $\bar{\sigma}^{s} = -\frac{E}{3(1-2\upsilon)}\mathbf{I} + \frac{E}{2(1+\upsilon)}\int_{0}^{t} G(t-\tau)\frac{\partial e^{s}}{\partial \tau}d\tau$ $G(t) = 1 + \bar{G}e^{-t/\tau_{\upsilon}}$	• Consider porosity, elasticity, and viscous nature of food matrix

In the existing literature, theoretical models have been developed based on some simplified assumptions; for example, treating food materials as rubbery [40] or elastic [46].

These assumptions simplify the problem formulation, but they are not conducive to a realistic understanding. Simplistic assumptions have also been made regarding the behaviour of fruits and vegetables. We will not do into to provide the detail on the models; interested readers are referred to existing literature. In the following section, the current challenges which are encountered during developing deformation models are discussed.

10.4 CHALLENGES

Both empirical and theoretical models have their limitations. Empirical models are unable to provide the physical understanding behind the defamation and have minimum flexibility. A specific empirical model only applies to the same material and the same drying condition. Slight changes in materials and methods make the empirical model erroneous. On the other hand, theoretical models consider coupled transfer process with solid mechanics for predicting deformation, which involves consideration of several real-time material properties. However, the unavailability of real-time mechanical properties makes the models less accurate. Followings sections discuss the main challenges of deformation models.

10.4.1 Moisture Dependent on Internal Stress

The internal stress developed by the moisture gradient is the main facilitator deformation during drying. In other words, the stress generated by the moisture gradient is the dominant contributor to plant tissue shrinkage during drying. The deformation model needs the value of internal stress developed due to both the moisture and temperature gradient. Without the correct value of developed stress, the deformation prediction would be erroneous. Most of the literature assumes a linear relationship of water migration and induced stress that does not reflect the accurate value of developed stress. Therefore, extensive research is required to determine moisture content-dependent internal stress generation during drying.

10.4.2 Appropriate Material Model

There are many models available in the literature for porous deformable materials. However, it is difficult to adapt the right model for a selected porous material under deformation. It becomes further complicated once the materials encounter a phase-changing environment such as drying. The combined effect of porosity, viscosity, and elasticity of the materials determines the characteristics of deformation under the internal stress conditions. It is the first step of the researcher to select an appropriate material model for the subjected materials. As there are large numbers of material models available with their corresponding advantages and shortcomings, it is difficult to choose the right one for the material of interest.

10.4.3 Real-Time Mechanical Properties

Mechanical properties of porous materials, especially biological materials, vary with moisture content and temperature. It can be higher or lower than the fresh material at drying condition with different moisture content. However, most of the existing ones use mechanical properties of fresh materials. Mechanical properties of subjected materials at room temperature cannot be used in deformation modelling during drying. Phase change also contributes to altering the mechanical properties with different moisture content. For instance, Young's modulus (E) may start increasing once the sample reaches the glassy phase. Consideration of real-time mechanical parameters in the mechanistic model could provide a more accurate prediction of material deformation. Due to the unavailability of such experimental data on variable mechanical properties of the porous materials, application of the complex mechanistic models in the prediction of deformation is not an accurate strategy.

10.4.4 Coupling of Deformation and Transport Phenomena

Drying is considered simultaneous heat and mass transfer problem that refers to real-time change of thermal- and moisture-dependent parameters that need to considered in drying models. Incorporating deformation phenomena in drying models makes them more complicated as shown in Figure 10.4.

The coupling of multi-physics requires a great number of dependent variables, resulting in the requirement of high computational energy. In most cases, interrelated properties of materials associated with different physics are not available in the literature and the measurement of those is no easy task. Therefore, most of the

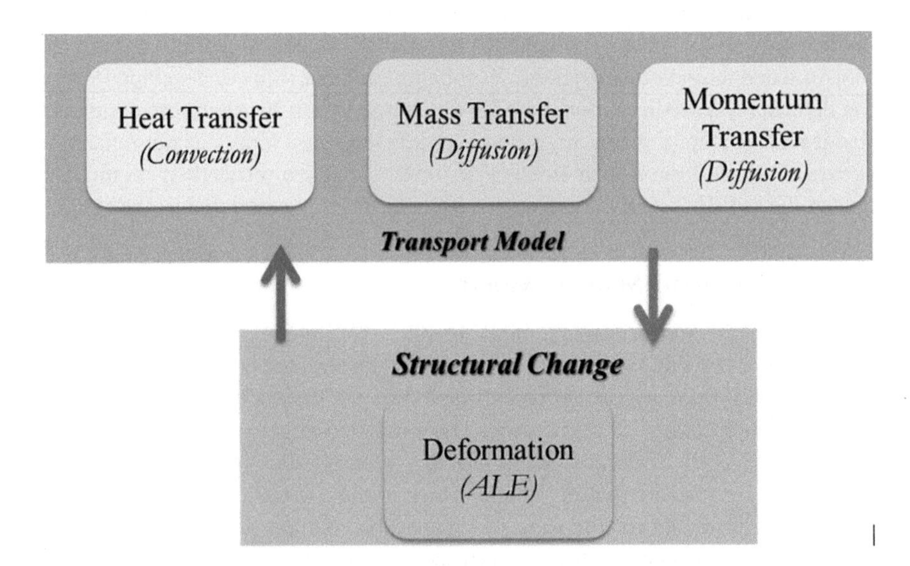

FIGURE 10.4 Coupled transport and deformation modelling during drying.

deformation models simplify the dependent properties and use constant values of different properties. Moreover, instantaneous changes in material properties due to different physics causes difficulties in synchronization during computation.

10.4.5 Multiscale Nature of Deformation

Deformation of porous material during drying is a multiscale problem. This means, the response of internal stress resulting in micro-level deformation and upscaling of micro-scale deformation eventually results in macro-scale deformation. Therefore, the proper understanding of the macro-level deformation of materials is critical for the prediction of deformation. In this sense, multiscale modelling can connect smaller and larger scales. Deploying multiscale drying modelling, deformation at different length scales can be predicted accurately for both biological and non-biological porous materials. In the literature, there are very limited studies dealing with deformation modelling using the multiscale model.

Physical quality of dried food materials largely depends on the extent of deformation during drying. Accurate prediction of deformation can contribute towards designing efficient drying process. The deformation of porous material depends on materials properties and process conditions. Empirical models can be developed with ease and offer high accuracy in prediction. However, empirical models are not higly recommended for diverse materials and drying conditions due to their lack of physical insight. By contrast, theoretical models are physics-based and offer more flexibility for diverse materials. However, a theoretical deformation prediction model is a complex task due to many challenges relating to instantaneous material properties and computational coupling strategies.

REFERENCES

1. M. I. H. Khan and M. A. Karim, "Cellular water distribution, transport, and its investigation methods for plant-based food material," *Food Res. Int.*, vol. 99, pp. 1–14, 2017
2. C. Kumar, M. U. H. Joardder, T. W. Farrell, G. J. Millar, and M. A. Karim, "Mathematical model for intermittent microwave convective drying of food materials," *Dry. Technol.*, vol. 34, no. 8, pp. 962–973, 2016.
3. S. Ghnimi, S. Umer, A. Karim, and A. Kamal-Eldin, "Date fruit (Phoenix dactylifera L.): An underutilized food seeking industrial valorization," *NFS J.*, vol. 6, pp. 1–10, 2017.
4. C. Kumar, M. U. H. Joardder, T. W. Farrell, and M. A. Karim, "Investigation of intermittent microwave convective drying (IMCD) of food materials by a coupled 3D electromagnetics and multiphase model," *Dry. Technol.*, vol. 36, no. 6, pp. 736–750, 2018.
5. M. Mahiuddin, M. I. H. Khan, N. Duc Pham, and M. A. Karim, "Development of fractional viscoelastic model for characterizing viscoelastic properties of food material during drying," *Food Biosci.*, vol. 23, pp. 45–53, 2018
6. J. E. Lozano, E. Rotstein, and M. J. Urbicain, "Shrinkage, porosity and bulk density of foodstuffs at changing moisture contents," *J. Food Sci.*, vol. 48, no. 5, pp. 1497–1502, 1983.

7. N. P. Zogzas, Z. B. Maroulis, and D. Marinos-Kouris, "Densities, shrinkage and porosity of some vegetables during air drying," *Dry. Technol.*, vol. 12, no. 7, pp. 1653–1666, 1994.
8. S. Khalloufi, C. Almeida-Rivera, and P. Bongers, "A theoretical model and its experimental validation to predict the porosity as a function of shrinkage and collapse phenomena during drying," *Food Res. Int.*, vol. 42, no. 8, pp. 1122–1130, 2009.
9. J. E. Lozano, E. Rotstein, and M. J. Urbicain, "Total porosity and open-pore porosity in the drying of fruits," *J. Food Sci.*, vol. 45, no. 5, pp. 1403–1407, 1980.
10. M. G. R. Perez and A. Calvelo, "Modeling the thermal conductivity of cooked meat," *J. Food Sci.*, vol. 49, no. 1, pp. 152–156, 1984.
11. R. S. Rapusas and R. H. Driscoll, "Thermophysical properties of fresh and dried white onion slices," *J. Food Eng.*, vol. 24, no. 2, pp. 149–164, 1995.
12. V. Rakesh, A. K. Datta, M. H. G. Amin, and L. D. Hall, "Heating uniformity and rates in a domestic microwave combination oven," *J. Food Process Eng.*, vol. 32, no. 3, pp. 398–424, 2009.
13. N. Duc Pham, M. I. H. Khan, M. U. H. Joardder, A. M. N. Abesinghe, and M. A. Karim, "Quality of plant-based food materials and its prediction during intermittent drying," *Crit. Rev. Food Sci. Nutr.*, vol. 59, no. 8, pp. 1197–1211, 2019
14. L. Slade, H. Levine, and D. S. Reid, "Beyond water activity: Recent advances based on an alternative approach to the assessment of food quality and safety," *Crit. Rev. Food Sci. Nutr.*, vol. 30, no. 2–3, pp. 115–360, 1991.
15. D. Champion, M. Le Meste, and D. Simatos, "Towards an improved understanding of glass transition and relaxations in foods: Molecular mobility in the glass transition range," *Trends food Sci. Technol.*, vol. 11, no. 2, pp. 41–55, 2000.
16. C. A. Angell, "Structural instability and relaxation in liquid and glassy phases near the fragile liquid limit," *J. Non. Cryst. Solids*, vol. 102, no. 1–3, pp. 205–221, 1988.
17. J. M. Del Valle, T. R. M. Cuadros, and J. M. Aguilera, "Glass transitions and shrinkage during drying and storage of osmosed apple pieces," *Food Res. Int.*, vol. 31, no. 3, pp. 191–204, 1998.
18. M. U. H. Joardder, C. Kumar, R. J. Brown, and M. A. Karim, "A micro-level investigation of the solid displacement method for porosity determination of dried food," *J. Food Eng.*, vol. 166, pp. 156–164, 2015
19. S. Achanta and M. R. Okos, "Predicting the quality of dehydrated foods and biopolymers—Research needs and opportunities," *Dry. Technol.*, vol. 14, no. 6, pp. 1329–1368, 1996.
20. V. Karathanos, "Collapse of structure during drying of celery," *Dry. Technol.*, vol. 11, no. 5, pp. 1005–1023, 1993.
21. M. E. Katekawa and M. A. Silva, "On the influence of glass transition on shrinkage in convective drying of fruits: A case study of banana drying," *Dry. Technol.*, vol. 25, no. 10, pp. 1659–1666, 2007.
22. L. E. Kurozawa, M. D. Hubinger, and K. J. Park, "Glass transition phenomenon on shrinkage of papaya during convective drying," *J. Food Eng.*, vol. 108, no. 1, pp. 43–50, 2012.
23. N. Wang and J. G. Brennan, "Changes in structure, density and porosity of potato during dehydration," *J. Food Eng.*, vol. 24, no. 1, pp. 61–76, 1995.
24. M. U. H. Joardder, A. Karim, C. Kumar, and R. J. Brown, "Pore formation and evolution during drying," in *Porosity*, Springer, 2016, pp. 15–23.
25. K. M. Waananen and M. R. Okos, "Effect of porosity on moisture diffusion during drying of pasta," *J. Food Eng.*, vol. 28, no. 2, pp. 121–137, 1996.

26. M. M. Rahman, M. U. H. Joardder, and A. Karim, "Non-destructive investigation of cellular level moisture distribution and morphological changes during drying of a plant-based food material," *Biosyst. Eng.*, vol. 169, pp. 126–138, 2018
27. J. A. Martínez-Casasnovas, C. Antón-Fernández, and M. C. Ramos, "Sediment production in large gullies of the Mediterranean area (NE Spain) from high-resolution digital elevation models and geographical information systems analysis," *Earth Surf. Process. Landforms J. Br. Geomorphol. Res. Gr.*, vol. 28, no. 5, pp. 443–456, 2003.
28. P. W. Kilpatrick, E. Lowe, and W. B. Van Arsdel, "Tunnel dehydrators for fruit and vegetables," *Adv. Food Res.*, vol. 6, no. 360, pp. 60123–60126, 1955.
29. S. Bhatnagar and M. A. Hanna, "Modification of microstructure of starch extruded with selected lipids," *Starch-Stärke*, vol. 49, no. 1, pp. 12–20, 1997.
30. R. J. Hutchinson, G. D. E. Siodlak, and A. C. Smith, "Influence of processing variables on the mechanical properties of extruded maize," *J. Mater. Sci.*, vol. 22, no. 11, pp. 3956–3962, 1987.
31. C. T. Huang and J. T. Clayton, "Relationships between mechanical properties and microstructure of porous foods: Part I. A review," *Eng. Food*, vol. 1, pp. 352–360, 1990.
32. P. S. Madamba, R. H. Driscoll, and K. A. Buckle, "Shrinkage, density and porosity of garlic during drying," *J. Food Eng.*, vol. 23, no. 3, pp. 309–319, 1994.
33. M. U. H. Joardder and M. A. Karim, "Development of a porosity prediction model based on shrinkage velocity and glass transition temperature," *Dry. Technol.*, vol. 37, no. 15, pp. 1988–2004, 2019.
34. M. S. Rahman, C. O. Perera, X. D. Chen, R. H. Driscoll, and P. L. Potluri, "Density, shrinkage and porosity of calamari mantle meat during air drying in a cabinet dryer as a function of water content," *J. Food Eng.*, vol. 30, no. 1–2, pp. 135–145, 1996.
35. J. Madiouli, J. Sghaier, J.-J. Orteu, L. Robert, D. Lecomte, and H. Sammouda, "Non-contact measurement of the shrinkage and calculation of porosity during the drying of banana," *Dry. Technol.*, vol. 29, no. 12, pp. 1358–1364, 2011.
36. M. S. Rahman, "A theoretical model to predict the formation of pores in foods during drying," *Int. J. Food Prop.*, vol. 6, no. 1, pp. 61–72, 2003.
37. P. Alvarez and P. Legues, "A semi-theoretical model for the drying of thumpson seedless grapes," *Dry. Technol.*, vol. 4, no. 1, pp. 1–17, 1986.
38. M. U. H. Joardder, C. Kumar, and M. A. Karim, "Prediction of porosity of food materials during drying: Current challenges and directions," *Crit. Rev. Food Sci. Nutr.*, vol. 58, no. 17, pp. 2896–2907, 2018.
39. R. May, G. Dandy, and H. Maier, "Review of input variable selection methods for artificial neural networks," *Artif. Neural Networks-Methodological Adv. Biomed. Appl.*, vol. 10, p. 16004, 2011.
40. T. Gulati and A. K. Datta, "Mechanistic understanding of case-hardening and texture development during drying of food materials," *J. Food Eng.*, vol. 166, pp. 119–138, 2015.
41. S. Jan Kowalski, "Mathematical modelling of shrinkage during drying," *Dry. Technol.*, vol. 14, no. 2, pp. 307–331, 1996.
42. N. Sakai, H. Yang, and M. Watanabe, "Theoretical analysis of the shrinkage deformation in viscoelastic food during drying," *Jpn. J. Food Eng.*,vol. 3, pp. 105–112, 2002.
43. M. Mahiuddin, M. I. H. Khan, C. Kumar, M. M. Rahman, and M. A. Karim, "Shrinkage of food materials during drying: Current status and challenges," *Compr. Rev. Food Sci. Food Saf.*, vol. 17, no. 5, pp. 1113–1126, 2018.

44. M. I. H. Khan, R. M. Wellard, S. A. Nagy, M. U. H. Joardder, and M. A. Karim, "Investigation of bound and free water in plant-based food material using NMR T2 relaxometry," *Innov. food Sci. Emerg. Technol.*, vol. 38, pp. 252–261, 2016.
45. Y. Llave, K. Takemori, M. Fukuoka, T. Takemori, H. Tomita, and N. Sakai, "Mathematical modeling of shrinkage deformation in eggplant undergoing simultaneous heat and mass transfer during convection-oven roasting," *J. Food Eng.*, vol. 178, pp. 124–136, 2016.
46. C. Niamnuy, S. Devahastin, S. Soponronnarit, and G. S. V. Raghavan, "Modeling coupled transport phenomena and mechanical deformation of shrimp during drying in a jet spouted bed dryer," *Chem. Eng. Sci.*, vol. 63, no. 22, pp. 5503–5512, 2008.

11 Multiscale Drying Modelling Approaches

11.1 INTRODUCTION

Most of the materials, especially the heterogeneous materials, are multiscale in nature. For example, a piece of plant-based material is made of an electron at a lower scale and described by their geometric dimension at a higher scale [1]. Similar to length scale, materials observation depends on the temporal scale. Magnifying of materials allows observing different characteristics of the material not available at their visible range. Similar to this, slowing down or hastening lime lapsing provides more information than those at normal speed. Like the property variation on a different scale, kinetics of physical phenomena also vary on a different scale. Due to this, drying phenomena can be treated as a multiscale problem [2].

Continuum-based macroscale models are easy to formulate and take less computational cost. However, the required apparent thermo-physical properties implicitly depend on the micro-level structural orientation of the material. Therefore, incorporation of micro-level properties is incumbent for accurate prediction. While microscale models can capture small-scale properties, they are computationally expensive in terms of cost and time [3]. Therefore, an optimized model of a physical phenomenon of multiscale nature cannot be dealt comprehensively with a single-scale modelling approach including micro-scale or macro-scale, as shown in Figure 11.1. Such a problem needs to be dealt with a multiscale modelling paradigm.

The multiscale modelling approach is well established in many branches of science and engineering; however, it is in its preliminary stages in the drying arena. A multiscale model is a hierarchy of sub-models that describe the material properties and physical phenomena at different scales in an interconnected way. These models are bridged in upscaling and downscaling approach according to the requirements discussed in the later sections of this chapter. Prior to the discussion on the multiscale modelling details, we need to delve into some very important issues.

11.1.1 The Hierarchical Structure of Materials

The structure of materials at different scales varies due to the diverse structural characteristics of different materials. The diversification is mainly caused due to the variation of compositions, interaction among molecules, presence of

DOI: 10.1201/9780429461040-11

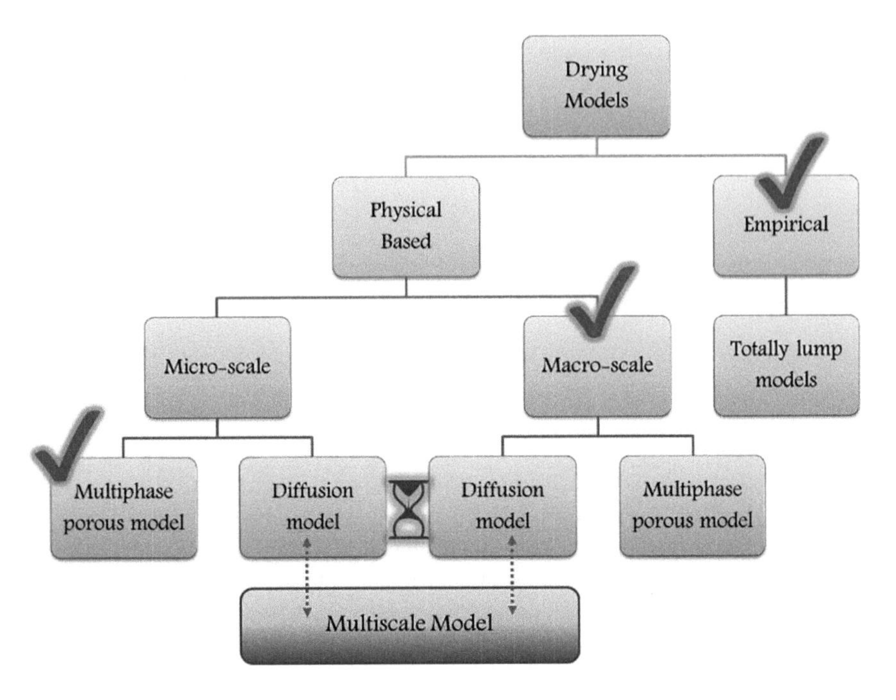

FIGURE 11.1 Multiscale modelling approach in drying phenomena.

porosity, and pore distribution [4]. Therefore, a generalization of the structure is not possible. However, classification based on similar composition, microstructural nature, and air space patterns can be achieved. Properties of materials and kinetics of physical phenomena vary with the variation of scales. Properties at smaller level influence the properties at higher levels, as depicted in Figure 11.2. It can be seen from the figure that different levels of diversity of component dominate the structural attribution of food materials. The similar hierarchical structure is available for most of the materials that are subject to drying phenomena.

The composition of materials plays a remarkable role in the forming of structure, maintaining physiochemical stability, and retention of nutrition. For example, the structure of food materials made of polysaccharides significantly varies from that of foods made from protein. Besides water content, its distribution contributes to the development of the structure of porous water-rich materials.

Therefore, it is commonly observed that loss of water or its absorption affects the structural features of porous materials. Water has a non-uniform spatial distribution inside porous materials and exhibits significant variations in the different moisture-dependent properties. Heterogeneous nature of different materials is the consequence of the presence of bulk, capillary, physically bound, and chemically bound water at different proportions [5].

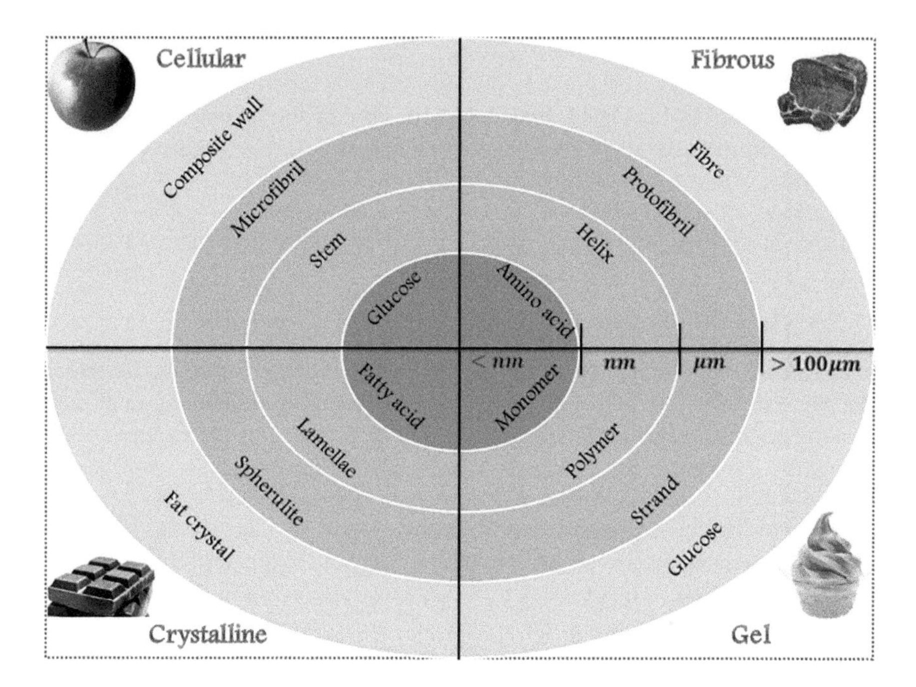

FIGURE 11.2 Conceptual multiscale system [4].

11.1.2 Structure–Properties–Drying Kinetics Relationship

Proper knowledge of structural changes during drying is essential to develop more rational and energy-efficient food drying. Retaining the structure of the wet porous materials and structural functionality is the utmost importance for the drying industries. Mass transfer in porous materials involves many different internal moisture transfer mechanisms that noticeably depend on the composition, properties, and initial structure of the drying process conditions.

Sample structures at different scales significantly affect transport kinetics, energy consumption, and quality enhancement, including deformation. For instance, the use of rugged surfaces instead of the plain ones gives more surface area to transfer the heat. Apart from this, due to the rugged shape, the airflow over the sample develops turbulence that results in higher heat transfer. Taking all these into consideration, waviness of the sample surface contributes to saving a significant amount of energy. Lower deformation in the rugged sample may be linked to the distribution of stress developed as the result of temperature and moisture gradient within the sample during drying. As the rugged surface samples demonstrate more rigidity than plane ones, the moisture migration causes lower structural modification during drying. Similarly, micro-level collapse is an important phenomenon that significantly affects the degree of deformation. After the removal of most of the water, pore collapse occurs at the latter stages of drying.

Despite the great importance of structure–function relationship, there are very few studies on the relationship between effective moisture diffusivity and change of structure during drying of food materials. One of the main hurdles in connection with the establishment of this relationship is the in situ determination of porosity. Microstructural image acquisition has high potential to correlate the drying kinetics and structural modification. Proper quantification of food microstructure can provide important insight into drying-related parameters.

11.1.3 Imaging: Structure and Properties Quantification

In multiscale modelling, one scale provides the properties of the other. Therefore, the properties of materials at different scales is crucial in developing multiscale modelling schemes. Observation of properties at different levels requires scale-compatible device. A wide variety of image acquisition techniques are available to to use for high-resolution images of food materials, including SEM, X-ray microtomography, confocal laser scanning microscopy, optical coherence tomography, and magnetic resonance imaging. Moreover, many other apparatuses are used in attaining properties of materials at different scales, as shown in Figure 11.3.

In multiscale, properties at the microscale are generally measured and used at the macroscale transport model using upscaling of the properties. Microscale diffusivity and permeability can be determined experimentally using nuclear magnetic resonance (NMR) and X-ray μCT. Moisture distribution throughout the sample can be visualized using X-ray μCT. Similar to mass transfer related properties, thermal properties at the microscale can be determined at a smaller scale. Scanning thermal microscopy (SThM) and atomic force microscope (AFM) are two suitable options for imaging and mapping the thermal conductivity variations across the sample at the micro-level. While specific heat of the corresponding material at the micro-level can be determined using differential scanning

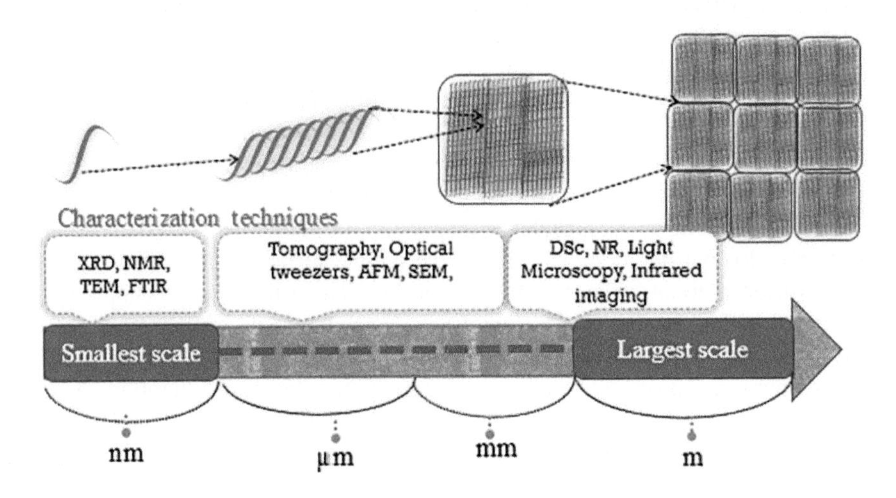

FIGURE 11.3 Property characterizations at different scales of materials.

calorimeter (DSC), mechanical properties including Young's modulus can be determined at microscale using nano-indentation using AFM.

All of the abovementioned imaging and properties measurement techniques at smaller scale have advantages and shortcomings. In the case of individual imaging of high resolution, there is a limited field of view, and these are eventually applicable for small sample sizes. To overcome this problem, "Multiscale Imaging" techniques have recently emerged in different fields. In multiscale imaging, special statistical methods are deployed to obtain high-definition images for larger samples.

11.2 MODELLING AT A DIFFERENT SCALE

Mathematical model at different scales is governed by different scale-compatible physics, and their corresponding modelling differ as well. In general, two broad categories of the modelling approach are available for modelling physical phenomena, namely continuum and discrete approaches. Continuum approaches use generally partially differential equations to govern the underlying physics using Finite element method (FEM), finite volume method (FVM), or finite deference method (FDM) for analysis. On the other hand, discrete models use statistic mechanics method such as discrete element method (DEM). Both problem formulation and computational solution differ in those approaches, as shown in Figure 11.4, which demonstrates the general computation paradigms for both approaches.

Depending on the degree of approximation, these simulation approaches can also be categorized as deterministic or stochastic. Modelling at different scales offers different advantages with their distinct shortcomings. In general, modelling at a smaller scale offers higher accuracy than their larger-scale counterparts. However, the larger-scale model takes less time and computation cost than smaller-scale modelling. In Table 11.1, modelling approaches at different scales have been summarized with their corresponding obtained information.

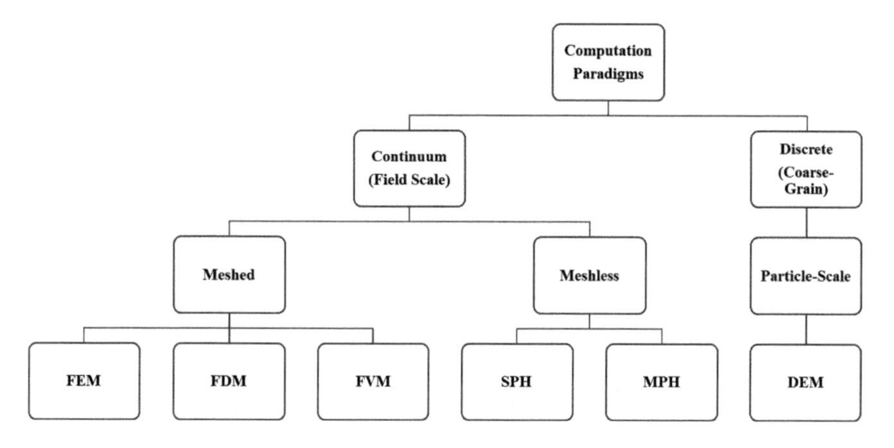

FIGURE 11.4 Computation paradigm at a different scale.

TABLE 11.1
Modelling Approaches at Different Scales

Scale	Modelling Approach	Solvent	Objectives
Quantum	Quantum mechanics	Explicit	Atoms, electron
Atomistic modelling	Molecular dynamics Coarse-grained	Explicit	Atomistic information
Microscale	Phase/force field modelling, Lattice-Boltzmann, Brownian Dynamics	Explicit/ implicit	Interfacial properties
Mesoscopic modelling	P-X-FEM Kinematic Monte-Carlo	Implicit	Phase (fluid) properties
Macroscopic modelling (Continuum)	FEM(PDEs)	Implicit	Transport phenomena

Different scales deal with different physics and retrieve information of various nature. Range of scale and related observation often vary in different arenas of engineering. For example, the modelling approach of micro and mesoscale are often confusing as different fields define and deal with these two scales differently. Even a combination of these two scales into a single scale can be observed in the literature. Therefore, the modelling approaches mentioned under these two scales can be interchangeable according to the definition of the scale.

In this section, we will only point out the modelling/calculation methods at different scales, and exclude the continuum approaches as these were discussed in Chapter 4. The reader interested in more insight into these topics can consult numerous other sources available, including the referred ones in the reference section.

11.2.1 Atomic Scale Simulation

The goal of the atomistic model is to attain the information of each atom of material from different perspectives. The compiled information of atoms allows us to understand the nature of phase changes and other phenomena. The data of this modelling approach provides the links between atomic scales to micro-scale phenomena.

11.2.2 Molecular Dynamics Simulations

Molecular dynamics mechanics (MDM) methods apply the laws of classical physics to predict the structures and properties of molecules. Classical molecular dynamics calculates the behaviour of molecules using the integrated equation of motion (Newton's, Hamilton's, or Lagrange's equations of motion) of individual molecules. Despite the inherited simplicity of MDM methods, the computation of millions of molecules makes it difficult in application. Therefore, reduction of the number of the equation of motions is often done using coarse-grain assumption.

11.2.2.1 Monte Carlo Method

Monte Carlo method (MCM) is a stochastic approach that can overcome some of the shortcomings of the molecular dynamics (MD). Unlike the MD method, MCM takes consideration of random atoms instead of taking the integration of all atoms' properties. MCM can model dynamic properties including diffusion of materials at atomic and micro-level. Therefore, MCM is found in modelling a wide variety of materials and physical phenomena in different field of science.

11.2.2.2 Coarse-Grained Models

Instead of taking individual atoms or molecule, a whole group of those atoms or molecules is considered as a single entity in the coarse-grain model (CGM). Reduction of the number of atoms leads to a high degree of freedom achieved in the CGM. CGM can be used in investigating the dynamics at longer length scale.

11.2.2.3 Lattice-Boltzmann Method

For modelling microscale and mesoscale models, the Lattice-Boltzmann method is useful. In the Lattice-Boltzmann method, materials and fluids are treated as quasi-particles that maintain the conservation laws along with the interaction of collision. In this model, the collisions-related calculations are accomplished using a discretized Boltzmann equation [6]. As this modelling approach uses a discrete element approach, it can handle complex geometries with moving boundary concepts.

11.3 MULTISCALE MODELLING APPROACHES: BRIDGING BETWEEN SCALES

The multiscale modelling approach is developed with appropriate bridging scheme among the models of different scale [7]. The main challenges in multiscale modelling are identifying the correlation among the models of different length scales [8]. Coupling the models of different scales need to be performed in a two-way approach, namely up scaling and down scaling. Homogenization or coarse-graining are the common terms that are referred to as the up scaling approach.

Localization and fine graining are often used to refer to down scaling methods. Whatever the terms used in coupling in multiscale modelling, the main purpose of inter-scale coupling is to utilize information of a scale to the model of the higher or lower length-scale model as shown in Figure 11.5.

In multiscale drying models, microscale and macroscale observation requires special attention. Models of other length scales, including atomic or molecule scale models, are often omitted to avoid modelling complexity. In a micro–macro scale drying model, upscaling of properties of microscale can be done using an appropriate homogenization technique [9]. For example, effective properties $\left(X_{eff}\right)$ of transport phenomena including effective moisture diffusivity (D_{eff}) and effective thermal conductivity (K_{eff}) can be obtained using one of the homogenization techniques.

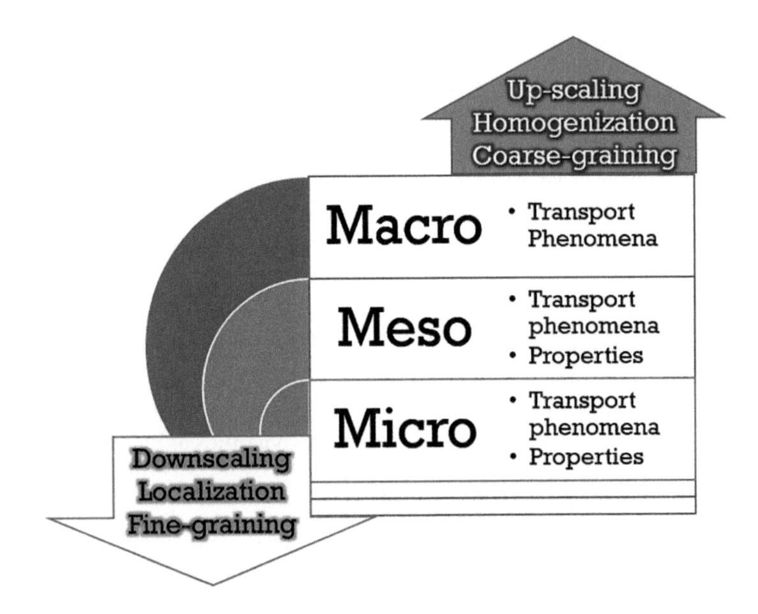

FIGURE 11.5 Bridging approaches between models of different scales.

These effective properties are then used in macroscale modelling as shown in the following equation where the effects of the governing equation are obtained from the homogenization of microscale modelling. The resulting energy balance equation can be written as:

$$\frac{1}{r}\frac{\partial}{\partial r}\left(rk_{eff}\frac{\partial T}{\partial r}\right)+\dot{g}=\rho_{eff}C_{P_{eff}}\frac{\partial T}{\partial t} \tag{11.1}$$

Formulation of the multiscale model and solution techniques are two different aspects of the model. Due to their equal importance, multiple classifications of multiscale are available in the literature. Based on the problem formulation approach and solution strategy, multiscale modelling approach can be classified as shown in Figure 11.6.

11.3.1 Problem Formulation

Based on problem formulation approaches, multiscale model paradigms can be categorized as concurrent, hierarchical, and hybrid models.

11.3.1.1 Concurrent Approach

When both of the scales are addressed simultaneously during the multiscale model formulation, it is classified as a concurrent approach. In this modelling approach, different modelling including molecular dynamics and finite element are applied in different parts of a single domain. For example, deploying of DEM can be possible in a continuum domain where attaining a solution using FEM encounters

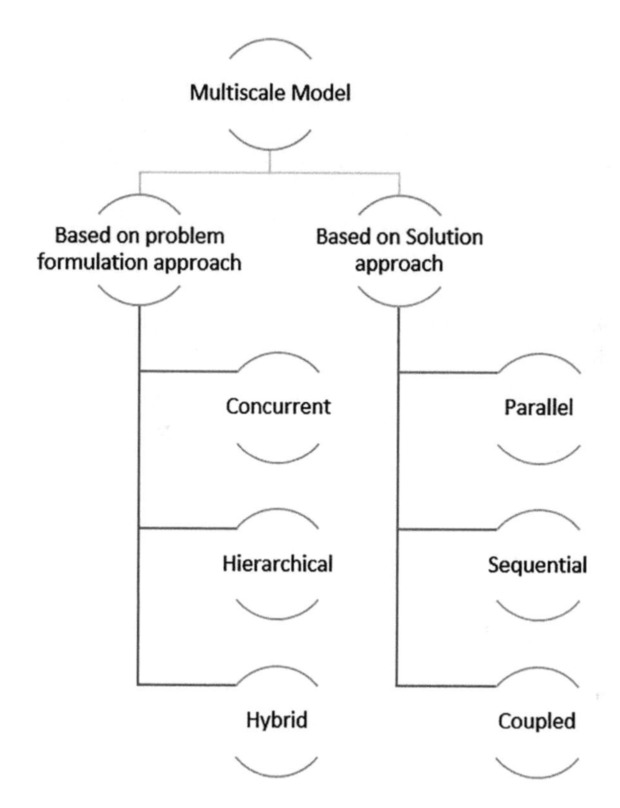

FIGURE 11.6 Classification of multiscale modelling approaches.

difficulties. Despite these advantages, the concurrent modelling approach needs significant dealing of boundary conditions between two heterogeneous modelling approaches, namely DEM and FEM. Moreover, determining and maintaining the pattern of growth of DEM in a continuum domain is still a challenge.

11.3.1.2 Hierarchical Approach

Hierarchical methods hierarchically link the scales. In other words, separate simulation of individual scales executed separately and passes as input to the higher- scale model. Volume average of variables s is the linking pathway in hierarchical modelling [10,11]. The hierarchical model is more suitable in transferring information during homogenization, whereas it is less effective in localization.

11.3.1.3 Hybrid Multiscale Modelling

A hybrid multiscale model is developed in the literature using continuum and discrete models [12,13]. This type of multiscale model provides the advantages of both continua such as finite element method and discrete model [14]. Hybrid methods also reveal the properties of quasi-continuum methods and wavelet-based methods [15]. Even the two types of modelling approach in continuum, namely finite element method and finite volume method, can be applied in the hybrid multiscale model [16].

11.3.2 Solution Approach

Similar to the multiscale model formulation, the solution of those models varies with the nature of solution steps associated with individual models. In parallel multiscale solution approach, models of different scale are computed simultaneously. Parallel computation of multiscale models is a promising option for high computation demand [17] along with ensuring modelling accuracy [18].

Parallel methods solve models of both scales separately. The sequential model is often suitable for passing a few numbers of parameters between the scales. In this approach, preprocessed parameters at one scale are used to characterize the model of another scale [19]. However, sequential coupling is difficult when the interrelated models depend on many variables.

Individual types of formulation and solution approaches of the multiscale model offer advantages and shortcomings when compared with other approaches. The appropriate approach needs to be selected based on scale definition and solution capacity for modelling a particular material drying.

11.4 CHALLENGES IN CURRENT MULTISCALE PARADIGMS

Though multiscale modelling has great potential in the modelling of drying phenomena comprehensively, it is relatively less practiced in this field [20]. Some challenges put constraints in implementing multiscale modelling in drying phenomena. The following are some remarkable challenges prevalent in multiscale modelling:

- Coupling of models at different scale is the main challenges due to numerous interdependent variables associated in the underlying multiphysics during drying.
- Dealing with coupled multi-physics of drying is a great challenge in formulating multiscale modelling. Simultaneous heat and mass transfer facilitate multiple mechanisms of moisture transport as well as changing phases. Multiple physics are coupled in drying, which makes the multiscale model considering all underlying physics more complicated.
- Real-time structural properties at different scales is essential for successful computation of multiscale drying model. Simultaneous heat and mass transfer along with deformation make the structure–property relationship of materials complicated.
- The heterogeneous domain of different materials is another constrain in formulating and computing multiscale modelling [21]. Therefore, the researcher often takes the homogenous materials for multiscale modelling, which eventually provide erroneous results.
- Selecting the nature of a modelling approach (continuum or discrete) for a scale of interest is often challenging. In addition, the separation of scale with appropriate characteristic length is still a challenging task.

- Determination of required resolution in multiscale modelling is important in the accuracy of the modelling as well as in the speed of computation. Choosing the smallest scale is a challenging task. The smaller the scale, the more properties will affect the drying phenomena as well as increase the computation time. Therefore, the selection of the finest scale of the multi-scale model needs special attention.

11.5 PROSPECTS: MULTISCALE MODELLING–ARTIFICIAL INTELLIGENCE INTEGRATION

Relatively new concepts of data management using machine learning are powerful techniques to couple multimodal and multiscale data and establish relations between multi-physics governed phenomena including drying [22].

Machine learning (ML) techniques can manage information from the acquired images/data of different scales [23]. Machine learning takes deep learning as the tools to train machines to classify and analyse raw and noisy data as shown in Figure 11.7. Incorporation of such trained machines in drying systems can provide real-time required data multiscale modelling and validation [24].

There are many modelling approaches to train machines that can deal with the enormous amount of information. Artificial neural network (ANN), fractal analysis, and fuzzy logic are some of the examples of machine learning models.

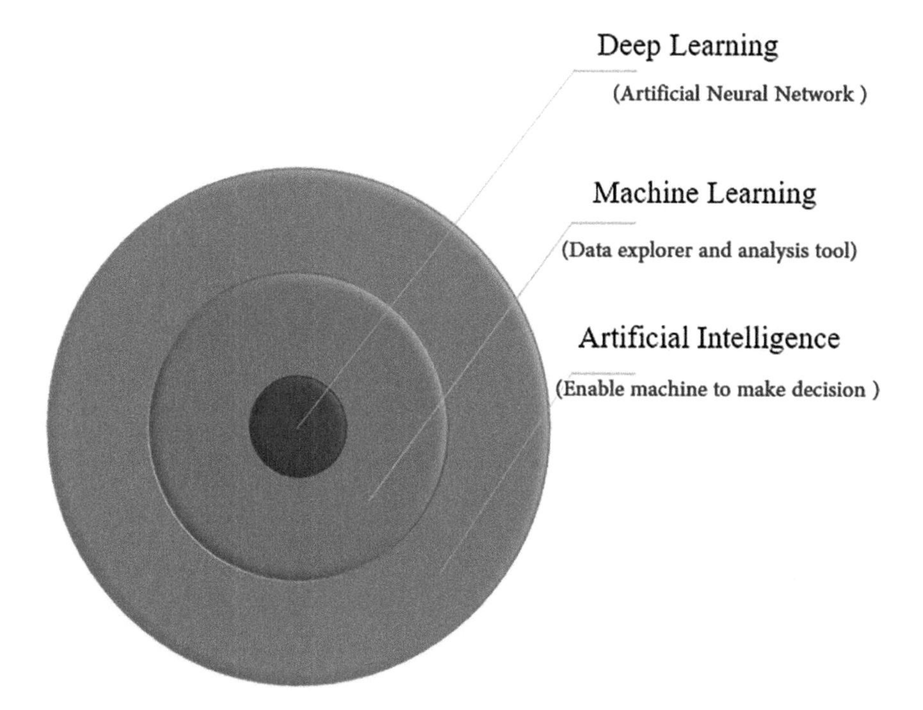

FIGURE 11.7 Artificial intelligence in multiscale modelling.

Real-time measurement of different qualities from food microstructure can be possible using fuzzy logic [25,26]. Fuzzy logic can be applied with simple mathematical relationships, which can be executed with commercial tools.

ANNs are mathematical models to collect and catalogue information presented in a given raw data from a sensor or probe [27]. The enormous number of neurons along with the vast number of connections between them are the foundation of ANN model development. Several layers lined up with neurons forms a multilayer perception network. Input data is fed in the first layer; whereas, the last layer provides the output of the model. Between the first and last layers, there are many internal hidden layers.

Fractal dimension is a geometrical concept and refers to the shape of an object that appears the same in different scale of observation. It can be used in stamping structural property information and provides quantitative data from the raw data analysis.

Several studies are available in connection with artificial intelligence and multiscale modelling in many fields, but there is still limited study regarding integrating these two powerful concepts of data analysis and prediction model for physical phenomena such as drying.

Therefore, researchers and scientists from different fields can incorporate their understanding and attempt to establish a bridging relationship between artificial intelligence and multiscale modelling to overcome the current challenge in drying models.

REFERENCES

1. A. Vigliotti and D. Pasini, "Mechanical properties of hierarchical lattices," *Mech. Mater.*, vol. 62, pp. 32–43, 2013.
2. T. Defraeye, "Advanced computational modelling for drying processes—A review," *Appl. Energy*, vol. 131, pp. 323–344, 2014.
3. Z. Welsh, M. J. Simpson, M. I. H. Khan, and M. A. Karim, "Multiscale modeling for food drying: State of the art," *Compr. Rev. Food Sci. Food Saf.*, vol. 17, no. 5, pp. 1293–1308, 2018.
4. M. U. H. Joardder, C. Kumar, and M. A. Karim, "Food structure: Its formation and relationships with other properties," *Crit. Rev. Food Sci. Nutr.*, vol. 57, no. 6, pp. 1190–1205, 2017.
5. M. U. H. Joardder, R. J. Brown, C. Kumar, and M. A. Karim, "Effect of cell wall properties on porosity and shrinkage of dried apple," *Int. J. Food Prop.*, vol. 18, no. 10, pp. 2327–2337, 2015.
6. R. G. M. Van der Sman and A. J. Van der Goot, "The science of food structuring," *Soft Matter*, vol. 5, no. 3, pp. 501–510, 2009.
7. B. Chopard, J. Borgdorff, and A. G. Hoekstra, "A framework for multi-scale modelling," *Philos. Trans. R. Soc. A Math. Phys. Eng. Sci.*, vol. 372, no. 2021, p. 20130378, 2014.
8. M. I. H. Khan, Z. Welsh, Y. Gu, M. A. Karim, & B. Bhandari, "Modelling of simultaneous heat and mass transfer considering the spatial distribution of air velocity during intermittent microwave convective drying," *Int. J. Heat Mass Trans.*, vol. 153, p. 119668, 2020
9. Z. G. Welsh, M. I. H. Khan, and M. A. Karim, "Multiscale modeling for food drying: A homogenized diffusion approach," *J. Food Eng.*, vol. 292, p. 110252, 2020.

10. J.-L. Bouvard, D. K. Ward, D. Hossain, S. Nouranian, E. B. Marin, and M. F. Horstemeyer, "Review of hierarchical multiscale modeling to describe the mechanical behavior of amorphous polymers," *J. Eng. Mater. Technol.*, vol. 131, no. 4, pp. 041206–041221, 2009.
11. J. J. de Pablo and W. A. Curtin, "Multiscale modeling in advanced materials research: Challenges, novel methods, and emerging applications," *Mrs Bull.*, vol. 32, no. 11, pp. 905–911, 2007.
12. J. M. Osborne et al., "A hybrid approach to multi-scale modelling of cancer," *Philos. Trans. R. Soc. A Math. Phys. Eng. Sci.*, vol. 368, no. 1930, pp. 5013–5028, 2010.
13. A. R. Anderson, *Hybrid Multiscale Models of Cancer: An Invasion of Equations.* AACR, 2006.
14. X. Yang, X. Meng, Y.-H. Tang, Z. Guo, and G. E. Karniadakis, "A generalized hybrid multiscale modeling approach for flow and reactive transport in porous media," *AGUFM*, vol. 2017, pp. H14G-07, 2017.
15. E. B. Tadmor, M. Ortiz, and R. Phillips, "Quasicontinuum analysis of defects in solids," *Philos. Mag. A*, vol. 73, no. 6, pp. 1529–1563, 1996.
16. N. Castelletto, H. Hajibeygi, and H. A. Tchelepi, "Hybrid multiscale formulation for coupled flow and geomechanics," in *ECMOR XV-15th European Conference on the Mathematics of Oil Recovery*, 2016, p. cp–494.
17. P. Macioł and K. Michalik, "Parallelization of fine-scale computation in agile multiscale modelling methodology," *AIP Conf. Proc,*, 2016, vol. 1769, no. 1, p. 60009.
18. Y. Shang, "A parallel two-level finite element variational multiscale method for the Navier–Stokes equations," *Nonlinear Anal. Theory, Methods Appl.*, vol. 84, pp. 103–116, 2013.
19. Q. Zeng and Y. Qin, "Multiscale modelling of hybrid machining processes," in *Hybrid Machining: Theory, Methods, and Case Studies*, Academic Press, 2018, pp. 269–298.
20. M. M. Rahman, M. U. H. Joardder, M. I. H. Khan, N. D. Pham, and M. A. Karim, "Multi-scale model of food drying: Current status and challenges," *Crit. Rev. Food Sci. Nutr.*, vol. 58, no. 5, pp. 858–876, 2018.
21. J. Walpole, J. A. Papin, and S. M. Peirce, "Multiscale computational models of complex biological systems," *Annu. Rev. Biomed. Eng.*, vol. 15, pp. 137–154, 2013.
22. M. Alber et al., "Integrating machine learning and multiscale modeling—Perspectives, challenges, and opportunities in the biological, biomedical, and behavioral sciences," *NPJ Digit. Med.*, vol. 2, no. 1, pp. 1–11, 2019.
23. D. Lu and Q. Weng, "A survey of image classification methods and techniques for improving classification performance," *Int. J. Remote Sens.*, vol. 28, no. 5, pp. 823–870, 2007.
24. M. I. H. Khan, S. S. Sablani, M. U. H. Joardder, and M. A. Karim, "Application of machine learning-based approach in food drying: Opportunities and challenges," *Dry. Technol.*, pp. 1–17, 2020. DOI:10.1080/07373937.2020.1853152
25. M. I. Chacón, "Fuzzy logic for image processing: Definition and applications of a fuzzy image processing scheme," *Advanced Fuzzy Logic Technologies in Industrial* Applications, Ed. Y. Bai, H. Zhuang, and D. Wang. Springer, pp. 101–113, 2006.
26. Y. Bai, H. Zhuang, and D. Wang, *Advanced Fuzzy Logic Technologies in Industrial Applications.* Springer Science & Business Media, 2007.
27. V. H. C. de Albuquerque, P. C. Cortez, A. R. de Alexandria, W. M. Aguiar, and E. de M. Silva, "Image segmentation system for quantification of microstructures in metals using artificial neural networks," *Matéria (Rio Janeiro)*, vol. 12, no. 2, pp. 394–407, 2007.

Index